COLIN SPENCER'S
VEGETABLE BOOK

COLIN SPENCER'S

VEGETABLE BOOK

PHOTOGRAPHY BY LINDA BURGESS

CONRAN OCTOPUS

In memory of Conal Walsh

Throughout the book recipes are for four people unless otherwise stated.
Both metric and imperial quantities are given.
Use either all metric or all imperial, as the two are not necessarily interchangeable.

Project Management: Lewis Esson Publishing
Editors: Lewis Esson and Penny David
Art Direction: Mary Evans
Design: Meryl Lloyd
Design Assistant: Ian Muggeridge
Photography and Styling: Linda Burgess
Food Styling: Jane Suthering assisted by Emma Patmore
Production: Jill Macey

First published in 1995 by Conran Octopus Limited,
37 Shelton Street, London WC2H 9HN
Reprinted 1996

Cataloguing in Publication Data: a catalogue record for this book is available from the British Library

ISBN 1 85029 645 6

Typesetting by Ian Muggeridge, London
Printed and bound in Hong Kong

CONTENTS

INTRODUCTION

Over the last ten years or so we have learnt to value vegetables more highly, not only for their singular flavours but also for their nutritional value as an essential part of a healthy diet. At the same time, more and more new and unexpected vegetables have appeared in the market. Some look and sound bizarre; others look appealing, but both caution and orthodoxy are strong inhibitors to kitchen experiments. So these pages contain old stand-bys and new exotic vegetables; also all those vegetables that one has seen around for years, yet one has never got around to trying. As our horizons grow ever-further distant, as we hear and see other cultures eating their way enthusiastically through their own cuisines, the challenge to try new foods and recipes is ever-more compulsive. Let us then be adventurous. Do, I implore you, experiment with the strange and unfamiliar, for your mind, palate, and cuisine will all be enriched.

Because of air-freight and the demands of ethnic groups within our societies, the vegetable market now reflects world trade. Upon a display shelf or market stall one can now travel from Paris to Tokyo, from California to Israel, from Madrid to Sydney. Such a polyglot display can even be off-putting in its richness. What is a yam? After all, in the USA they call sweet potatoes 'yams', but surely they are different families altogether. The same vegetable is called an 'eddo' on one stall and 'dasheen' on another. And what on earth is 'elephant's ear'? The answers are all in these pages.

You will find here all the vegetables that we now commonly eat and which are regularly on sale in our markets. You will also find wild vegetables – like seakale and rock samphire – that were once popular but which we now sadly neglect.

LEFT Swiss chard

The vegetables in this book have been grouped broadly into their botanical families, eight of them in all, but with four smaller families grouped under the heading Tropical and Exotic Vegetables and a further chapter on Mushrooms & Truffles – making ten chapters in all. To find a particular vegetable, consult the index. If you know its family, you can go straight to the relevant chapter. A short history is given of each vegetable, followed by notes on its nutritional value, some information on the varieties, advice on buying and storage, guidance on its preparation, a summary of its general cooking qualities and finally some particular recipes.

Every vegetable has a history. Some, like taro, are so ancient that no wild variety exists today; others date from just before cultivation began; others are astonishingly recent – the Swede or the orange carrot, for example. All have a tale to tell; all have incidents in their lives which reflect on humankind, on what we thought and felt, exposing our vanity and aspirations, our most intimate personal habits and beliefs – as revealing as any archaeological remains. Why, for example, should we always think the consumption of garlic and onions is so vulgar? Why have so many vegetables been considered aphrodisiac? Especially rocket, the only vegetable the seeds of which were strewn around statues consecrated to Priapus. Did the Ancient World know something we don't? They certainly valued the medicinal quality of vegetables and herbs in a way which we have almost lost. How interesting, too, to discover so many of their techniques forgotten in the West but still carried on in the East. The Greeks used to wind-dry cabbage leaves and this practice is still continued in China today. Cabbage was greatly valued for many qualities, not least that it was said to protect against drunkenness.

ABOVE *Vegetables on sale in Madeira*

Our relationship with the plants that we have used for our survival is close and complicated and dates from the earliest beginnings of our evolution.

IN THE BEGINNING

There are few places upon the Earth where plant life is impossible – deserts of extreme aridity, mountainous peaks, regions of permanent ice – and these are the same regions where humans can barely exist. Even so, for a short period each year the desert flowers after rainfall and plants race through their life-cycle in that abrupt interval. While humankind, if passing, would also harvest what plant food it could gather.

We could not have survived without plants, for most of all they sheltered us from the elements. The tropical rain forests, the temperate deciduous forests, the grasslands, all act as blankets of vegetation that soften the impact of wind, of heat, of frost and of rain. Forest was shelter for humankind: it was dry, shady and cool. In colder climes, the forest was warmer – the forest was home. It was also the larder: so much of what we still eat now was eaten then. Mushrooms and fungi we still pick from the wild, but berries also – not just blackberries, but bilberries, wild strawberries and cranberries. However, I suspect nowadays we would be at a loss to feed ourselves outside the autumn season for so many of our cultivated vegetables we do not recognize in the wild. While many greatly sustaining wild foods we neglect for there is no necessity for us to forage.

In the early spring foods are limited. There is, however, a range of plants that store their nutrition over the winter for their coming growing season. For example, the onion family, plants of their close cousins the lily family, and the rhizomes of Solomon's Seal (*Polygonatum multiflorum*) which lie near the surface. The rhizomes of bracken can be dried and ground into a flour. Water plants were also a rich

ABOVE The vegetable market in Bali, Indonesia

source of flour: the tubers of the yellow water-lily (*Nuphar*), arrowhead (*Saggitaria*) and bulrush (*Schoenoplectus*) were the three species of lotus (*Nymphaea*) in Ancient Egypt which were all dried, ground and baked.

In early spring, as soon as the frost fades, the roots of the trees begin to take up water and then the sap starts to flow up the trunks. Sap is rich in sugars from the starch that his been stored over winter and this is a valuable food source. We are familiar with maple syrup, but other species of trees, such as birches, lime and aspen, all produce abundant sap.

A week or so later in spring, plants start to send up their shoots. These are all highly nutritious (as we know when we sprout seeds or beans), for they contain the growing energy of the plant. Asparagus is one of the few plants we still eat at this vital stage. Sea kale is another. However, in the forest many shoots would be blanched naturally beneath the thick covering of leaves on the forest floor. Bellwort, common pokeberry, sea holly, bracken, seminole asparagus and the asparagus bush are but a few whose growing shoots can be eaten. Also, baby pine cones and buds of trees such as lime, were a rich source of protein.

At this time of year deciduous forests contain an important food that we know now as slippery elm, a highly nutritious and digestible diet for invalids and people with gastric disorders – the food is cambium. In the spring, the trunks of trees begin to widen, due to a single layer of cells (the cambium) which ensheathes the trunk. These young cells are rich in protein and carbohydrates. The exterior bark of the tree would have been cut away and the new moist cambium scraped off. Such removal of living tissue would have certainly reduced the vigour of a tree, if not killed it if the cambium was scraped away in a ring around the tree. Slippery elm comes from one particular species – *Ulmus rubra* – but poplar, ash, lime and even pine, as well as elms, could have been used. Up to very recent times a bark bread was made in Scandinavia from cambium taken from an elm – *Ulmus glabra*.

In late spring and early summer, all kinds of new green leaves could be eaten – a huge range, when you consider there are 3,000 species of food plants from which humankind could forage. We know that the pounding of roots and leaves occurred before the discovery of fire. We might conjecture, then, that the mucilaginous plants – the leaves of mallow (*Malva*) being but one – would have been useful in holding disparate ingredients together. Certainly the mucilaginous plants (okra is one we still eat) were widely used. Just as seaweed was foraged by all peoples who lived near river estuaries and the sea.

Particular families of plants have always been favoured. The Umbelliferae or parsley family, which includes the wild carrot and parsnip, rock samphire and fennel. Another member never eaten now, the earth-nut or pig-nut (*Conopodium majus*) is a small edible tuber that has never been cultivated but always known to be delicious. In Europe it can be dug up from June onwards (see Richard Mabey's *Food for Free*). Another family is Polygonaceae, the buckwheat family, the seeds of which are nutritious and have always been eaten by animals and ourselves. Rhubarb and dock were eaten, and sorrel – with its astringent gooseberry flavour – must have delighted early foragers, as well as the fact that it shoots out of the ground so early in spring. There was another favourite in this family – the root and leaves of bistort (*Polygonum bistorta*), used until very recently in Yorkshire as the main ingredient in dock pudding – though this only uses the leaves. The root was soaked to steep out the tannin, then dried and ground into a flour, but the first young shoots were eaten as well.

Throughout spring and summer, flower-buds were gathered, especially the largest like the water-lily (*Nymphaea*). Once flowering is over, it is the seeds that become a source of food, as in the dandelion, the vetches and all the grasses. Berries and soft fruits are almost too obvious to mention, but one imagines their sweetness was welcomed for we have no idea how early the gathering of wild honey began.

When winter is threatening, the problem is to find foods that store – nuts are the obvious answer. Hazelnuts, walnuts, sweet chestnuts, pine nuts (*Pinus pinea* and *P. edulis*), butternuts (*Juglans cinerea*), hickories (*Carya*), beechnuts (*Fagus grandifolia*) and acorns – for the oak was the dominant tree of all the temperate forests. Roasting or boiling acorns destroys their bitterness and the tannin has to be extracted, so acorns could only appear as a winter staple after the discovery of fire. If the ground was not frozen hard, bulbs and tubers could be dug up. There is also evidence of winter storage of seeds and the drying of fruits, currants and wild strawberries. The fungi were also of immense value at this time, especially the large ones like *Fistulina hepatica*, a massive bracket fungus that grows on trees. We must not also forget lichens, like the reindeer moss (*Cladonia rangiferina*) which is one of the staple foods of reindeer and caribou. There are also algae growing over ponds and lakes; the Aztecs gathered blue-green algae (*Cyanophyta*), dried it and made it into cakes which stored well.

It has to be said, however, that gathering and foraging are haphazard. There is an enormous element of luck, and food sources can so easily be wiped out by floods, frosts and earthquakes, or by herds of grazing animals. How much better it would be for humankind if they could control the food they searched for, if it could be grown near their shelter. The impulse to settle and garden must have been there among small clans and individuals world-wide long before domestication actually began.

The other fascinating aspect of the early gatherers is how much experimentation was needed, for so many of the food families have toxic parts or members. Take the Solanaceae or potato family, how many people died from the berries of deadly nightshade? How many found out that tomato leaves and stems, potato leaves and the green patches on the potatoes themselves are poisonous? What of cassava, a staple food over much of the tropical world? How did people find it was edible once the cyanide had been washed out? The older members of any small group must have been greatly valued, for it was they with their experience who could memorize the edible parts of a plant and where it was likely to grow and at what time of year. The botanical knowledge must have been vast, passed on to each new generation: the plants that heal, the plants that refresh as drinks when steeped in water, as well as the plants that you ate. This knowledge could not be written down, language itself must then have first come about to describe the food that lay in and around the earth, its characteristics and how it was to be eaten.

This is a very crude summary of the potential food larder in wild plants which humankind needed to survive. It is crude because I have purposely omitted dates and places. I merely wanted to suggest to the general reader how the cultivated plants eaten now have their distant roots in the past and that, nutritionally, humankind might even have had a better diet then than we have today. Out of those 3,000 species of plants eaten for food, only 150 of them were ever cultivated and today the world lives off just 20 main crops. Most importantly, I want to stress how several million years of living from plants and then crustaceans and plants – long before hunting began – has a possible significant message for our present-day metabolism. Humankind has relied on wild foods for 99.8% of our time and through these millions of years human physiology has adapted to such foods. According to *The Driving Force – Food in Evolution* by Michael Crawford & David Marsh: 'There is a diet to which humans are adapted; this diet includes regular exposure to substances on which human metabolism is dependant, only some of which to date have been labelled as "essential nutrients".'

According to a scientific paper 'Vegetables, Fruit and Cancer' by Kristi A. Steinmetz and John D. Potter (*Cancer Causes and Control, Vol 2* 1991): 'vegetables and fruits contain the anti-carcinogenic cocktail to which we are adapted. We abandon it at our peril.'

I believe strongly that vegetables, grown organically, are vital to our health and spiritual well-being. Of course, they can be in a mixed diet of fish and meat, but the latter in my view should be an occasional garnish or flavouring.

ON NUTRITION

Understanding the chemical nature of our food is a very recent science. It began in the eighteenth and nineteenth centuries, when chemists isolated the main food components and divided them into carbohydrates, proteins and fats. The food components that occur in very small amounts took longer to analyse – the vitamins, pigments and flavour compounds required twentieth-century laboratory techniques before they could be isolated and characterized.

However, the greater our knowledge the greater it also clarifies how little we know. In fact, the newest research is inclined to give to the micro-nutrients – the trace minerals – and the way in which they interact with each other in a mysterious cocktail – more possible power in our metabolism than had ever seemed possible before.

So, though each of the entries in this book has some nutritional information, this cannot be absolute in any way. For example, tests on various apples for their vitamin C content found that it ranged from 3.1 mg per 100 g to 31.8 mg. This kind of range must be true of all the vegetables included in this book. Such a range cannot be otherwise, because plants are living entities as different in themselves as we are individuals. Consider, the nutrition of each fruit and vegetable is influenced by the soil and any fertilization it gets in its growth and these factors are a moveable feast. The nutritional facts given herein then are merely rough guides. For greater detail and an attempt at precision by taking the mean from a range of samples, consult *The Composition of Foods*, by McCance and Widdowson (5th edition MAFF).

The World Health Organization Committee wrote in the *WHO Report on Diet and Chronic Diseases* (1990), 'vegetables and fruits are low in energy, but high in fibre, vitamins and minerals. They play a protective role in preventing the development of cancers.' It is believed that vitamins C, E and beta-carotene are those which are involved in attacking and destroying the free radicals responsible for cancer. We should aim to eat at least 400 g / 14 oz of fresh fruit and vegetables per day – the equivalent of 4 apples. I try to eat a large mixed salad every day: simple in the summer when there is so much to choose from and such a variety of green leaves (remember that the darker the leaf, the more vitamins they contain); but winter salads of raw grated vegetables can be just as good and perhaps even more tasty.

Alliaceae

THE ONION FAMILY

ONION *Allium cepa*
GARLIC *Allium sativum*
LEEK *Allium porrum*
SHALLOT *Allium ascalonicum*
CHIVE *Allium schoenoprasum*
CHINESE CHIVE *Allium tuberosum*
WELSH ONION *Allium fistulosum*
ROCAMBOLE *Allium scorodoprasum*

Liliaceae

ASPARAGUS *Asparagus officinalis*

Until quite recently the onion and its close relatives were classified as members of the Liliaceae family – the lilies, rather than constituting a family of their own. They are also considered by many to form a subdivision of the Amaryllidaceae (which comprises the amaryllis, the narcissus and the snowdrop).

*The onion family itself has about 325 members, most of which seemed to originate in Asia, but seventy of which are indigenous to the Americas. Some varieties of onion grow wild, like the crow garlic (*Allium vineale*) or ramsons (*A. ursinum*), while many others are cultivated as vegetables.*

Except in the case of the leek and chive, it is generally the bulbs that are eaten. Some, like chives and Chinese chives, are popular as herbs and have been used as such for centuries. Extraordinarily enough, leek and garlic are cultivars not found in the wild, which shows how very ancient their culinary use is.

For our purposes we also include in this chapter the one member of the Liliaceae family cultivated as a vegetable – the asparagus.

ONION

Allium cepa

The onion and all its nearest relatives recede far back into prehistory. As well as other roots, like the wild turnip and radish, the wild onion must have been eaten with pleasure by man throughout his evolution, even if – before the advent of fire – chewing through uncooked fibrous matter must have been at times unrewarding. All the more exciting, then, to find a root which did not need a great deal of mastication and which enflamed the mouth, therefore stimulating the gastric juices, and which was also full of its own juice. Once the first grains were cultivated, the onion also grew well in the semi-humid climate found along the banks of irrigation canals.

Though its native country is unknown (it grows in all temperate zones), the onion is thought to be indigenous to the Middle East – the cradle of civilization. It is mentioned in the Bible (*Numbers* 11:5) as one of the foods the Israelites longed for in the wilderness. Certainly it was common fare in Egypt; the Greek historian Herodotus, writing in the fifth century BC, says that there was an inscription on the Great Pyramid stating the sum expended as 1,600 talents on onions, radishes and garlic which had been consumed by the labourers when building it.

Carvings of onions appear on the inner walls of the pyramids of Unas (*c*.2423 BC) and Pepill (*c*.2200 BC) in tombs of both the Old and New Kingdoms. Wooden models of onions have been found in tombs; as has evidence of the practice of placing onion bulbs in the bandages of mummies – in their armpits, eye sockets and body cavities – in an attempt, it is thought, to stimulate the dead to breathe again. The onion was also used as a means of assessing female fertility: it was introduced into the vagina and if the smell could be perceived next day in the mouth of the female it meant she was fertile; if not, the woman was sterile.

Some Egyptian priests refused to eat onions and garlic. This could be because they were thought to incite lust, a conviction that was to continue for some time, well into the Middle Ages. Another reason given was that the onion was the only plant to thrive throughout the waning of the moon. Though the taboo existed about consuming the onion, onions were not excluded from the altars – paintings frequently show priests holding them or depict an altar covered in the plants. A proverb emphasized another aspect of the onion of which priests might have been wary: 'If your mother is an onion and your father garlic, how could your smell be sweet, poor chap?'

Bread and onions were certainly the basic diet not only of Egypt but of Mesopotamia. As now, onions were preserved throughout the winter by being first dried, then plaited into strings, which is how they were sold then and up to fairly recent times.

An ancient myth of the Turks places onion and garlic at beginning of time, when the devil was sent out of Paradise and first set foot on Earth. On the spot where he placed his right foot there grew the onion and in the place of the left grew garlic. It is interesting to note the inference that both these strong-tasting bulbs – fiery and intense in their raw state – had for the peoples of the eastern Mediterranean a certain diabolism.

Onions were also popular in Ancient Greece: the physician and 'father of medicine', Hippocrates, mentions in 430 BC that onions were commonly eaten, other important vegetables being garlic, cabbage, peas and lentils. Theophrastus (370 BC), the philosopher and pupil of Aristotle, lists a number of varieties of onions and garlic in his botanical works.

All the varieties with which we are now familiar existed in classical times. The Roman writer Pliny (AD 79) claims that the round onion is the best, but that red onions are more highly flavoured than white. Palladius (AD 210) gives detailed directions for the culture of the onion. In Roman cooking, if we take the work of celebrated gourmet Apicius (AD 230) as an example, the onion was used to season dishes rather than as an ingredient. In his celebrated recipe book *De re coquinaria*, there are thirty-three recipes where the onion is used as a flavouring and only two dishes where it is used for itself: a purée of lettuce with onion and roasted meat served with onions. The Romans much preferred leeks. This, however, is a class difference: the poor began the day with bread and raw onion, and this is very likely why Apicius does not include the onion as a vegetable itself (except for the two recipes mentioned above), for to do so would be to add a common, even vulgar, note to meals for the ruling élite.

However, Roman distaste had some allies in the East; much of India refused to eat either onions or garlic. The higher castes, the Brahmins, refused on the grounds perhaps that onions represented common food, or that as they were allied with the cooking of meat they were somehow tainted. There is still suspicion about the use of onion in cooking; it still has class connotations, the raw onion is not eaten in polite society.

After the upper-class Roman aversion for the onion, it was the barbarians who brought the onion back into mainstream cooking, as they added it liberally to their roasting meats. Charlemagne ordered onions to be planted, and in centuries to come they frequently appeared in feudal tithes as one of the crops to be paid to the landowner.

Onions were indispensable in medieval cooking and were used in soups, broths and stews for, at this time, strong aromatic pungent flavouring was preferred. Onions cooked for a length of time with ginger, caraway, mustard, cardamom and coriander were favourite flavour combinations. They continued to be popular throughout the Renaissance, but the onions were diffused by cooking, their strong flavours part of the context of many a meat dish in unison with other herbs, spices and flavourings.

Much that modern research has shown to be efficacious about the onion was somehow absorbed by the ancient world, and folklore became full of the health-giving properties of both onion and garlic. In *The Four Seasons of the House of Cerruti*, a fourteenth-century Latin manuscript which is based on the wisdom of the medieval alchemists, the onion is praised as follows:

'An excellent thing, the onion, and highly suitable for old people and those with cold temperaments, owing to its nature, which is hot in the highest degree, sometimes moist, and sometimes dry. The most desirable of the many varieties are the white ones, being rich in watery juices. They generate milk in nursing mothers and fertile semen in men. They improve the eyesight, are softening, and stimulate the bladder. Headaches, which are sometimes caused by onions, can be cured with vinegar and milk. Those suffering from coughs, asthma and constrictions in the chest should eat boiled onions, or onions baked under the embers, served with sugar and a little fresh butter.'

Though the Old World's onion varieties were introduced into the Americas, the peoples there had their own indigenous onion types. Wild leeks have always grown in the woods of the North-east, and the American Indians liked wild garlic and chives. Bernal Díaz, the writer who accompanied the conquistador Cortés in Mexico, remarked that the Indian arrow shafts smelt strongly of garlic. In 1674 Father Marquette, the Jesuit explorer, told of being saved from starvation by eating the American native varieties (tree onions and nodding onions) on the southern shores of Lake Michigan. Marquette camped on what was to be the site of Chicago, a name which is Indian for the odour of onions. Oddly enough, though, the varieties later grown for cultivation in America were all brought over from Europe.

GROWING AND VARIETIES

There are 147 different varieties of *Allium cepa* on sale in Britain from seedsmen, so the British gardener has plenty of choice on this the most perennially popular of all vegetables. Firm favourites include: Ailsa Craig, round and mild: Stuttgarter Giant, with slightly flattened bulbs; and Sturon, which is preferred by growers as it has bolt resistance.

The cheapest way to grow your own onions is to sow 30 g / 1 oz of seed. It will germinate within three weeks. Once transplanted, onions will take twenty-two weeks to mature if started in spring and forty-six weeks if planted in autumn. You should harvest 3.5 kg/ 8 lb of onions from this amount of seeds. It is no good if you live in cold climes, unless you germinate the seed under glass. Onion seeds are also vulnerable to pests, particularly the onion fly, hence the popularity of onion sets. These are guaranteed to be virus-free and heat-treated so that they do not bolt. Onion or shallot sets are just the immature bulbs; once planted, they will take about twenty weeks to mature.

PICKLING ONIONS: For commercial pickling and adding to dry martinis, we use small white silverskin onions – obtained by sowing appropriate varieties deep in the soil. Paris Silverskin is lifted when the bulb is the size of a marble. Barletta is another similar variety. For home pickling, small onions are used from the onion varieties below.

RED ONIONS: These are now readily available in stores. They are milder and sweeter in taste and are best used raw in salads or added to the marinating liquor for a *ceviche*. There are other red varieties which you can grow: Long Red Florence is the torpedo-shaped red onion from Florence; Purplette are small onions which are good to use raw as they are so decorative but if pickled they turn pink.

SPANISH ONIONS: These are large varieties of yellow onion, about the size of a grapefruit, which are generally globe-shaped, sweet and mild. The white variety of Spanish onion has a flatter top and bottom than the original Spanish, but there is no difference in the flavour. Spanish onions are popular in the USA with a similar type – Bermuda. They used to be more popular than they are now – cookery books of forty years ago abound in recipes for stuffing and baking them. If baking, always first par-boil them for ten minutes. They are amazingly filling, whatever the stuffing.

SPRING ONION: On the whole these are not separate plants: they are merely young *Allium cepa* which are grown from seed and harvested while the green shoots are sturdy and before they die off – the thinnings, in fact. There are now, however, several varieties which are grown specifically as salad 'spring onions' or scallions (named after the ancient Palestine port of Ascalon). They are all white-skinned, but can be

RIGHT Red and yellow onions

highly pungent and peppery or mild. White Lisbon is the most popular of the varieties, but Ishikura, which has long straight stems and never forms bulbs, is also becoming readily available. All spring onions contain traces of protein, carbohydrate, fibre and very little fat and are excellent raw or briefly added to stir-fried vegetables at the end of cooking.

WHITE ONIONS: These are the types of onions which are on sale all the year round. They come in all shapes and sizes: round, oval, large and small. Their flavours can vary from mild to strong and there is no certain way of telling which will be which, except by trial and error.

YELLOW ONIONS: These come in two sizes: the larger are available all the year round from Poland and Spain; the smallest and dark-brown ones are the French variety, Jaune Paille des Vertus. Smaller yellow onions come from Japan and are onions used in pickling as well as in Oriental cuisine. The bulbs tend to be flattish and in some varieties, like Express Yellow, the flesh is clearly tinged with yellow. Others, like Imai Yellow and Senshyu, are paler. Yellow onions can be grown in late summer, but they do not keep well. There is little difference in flavour between yellow and white onions.

NUTRITION AND MEDICINAL

Both onion and garlic are packed with therapeutic compounds. The onion contains vitamins B, C and E, carotene, calcium, iron, phosphorous, potassium, sodium, sulphur and traces of copper. Both onion and garlic also contain a natural antiseptic oil, allyl disulphate and cycloallin. The latter is an anti-coagulant which helps to dissolve clots which form on the walls of blood vessels, hence onions and garlic do reduce the risk of heart disease. The really good news is that cooking – by boiling, frying or roasting – does not destroy these therapeutic properties. What is more, very little onion (only 10 grams – about a tablespoon) eaten per day will lower blood cholesterol.

CHOOSING AND STORING

Buy onions that are hard and firm with no sign of green shoots. Buy them in strings, if you can, and hang them up somewhere dry and cool. Onions are intended to keep throughout the winter and perhaps because we know this we tend to forget about them, finding them months later sprouted and shrivelled. So try and keep onions where you will notice them every day and test for freshness by pinching them. If there is any sign of softness, then use them up. The golden skins are an excellent colouring for stocks and hard-boiled eggs.

PREPARING

Once an onion is cut it releases a volatile substance (a sulphur compound) which makes the eyes water and then in turn produces sulphuric acid in the tears. No wonder it stings. These odour compounds are quickly driven off by cooking and some of them become converted to another complex molecule which can be 50–70 times sweeter than sugar. The reason for the onion's having so much juice is that the many layers are all leaf bases which surround very small shoots. These store water and starch during the bulb's first year of growth, for use during the second.

The best way to prepare an onion is as follows: first slice off the top and bottom. The coloured outer skin should then flake away readily with a little help from the knife or fingernail. As it is frequently dry or damaged, the layer beneath the skin often also needs to be removed. Occasionally you need the onion to be sliced fan-shaped but still held together by the basal stem (this can look good in dishes in which the onions are not cooked too long). To achieve this effect, cut the onions lengthwise, stopping short of the base. Otherwise, gouge out the basal core of the onion, slice the onion in half through the core and place each half cut side down. Holding it firmly with the fingertips of one hand, slice it across thinly for crescent parings as used raw in salads: for finely chopped dice, slice again lengthwise.

If shedding tears while preparing onions bothers you, try wearing goggles or cutting under water. The former is a good measure if preparing onions for pickling, which can take a long time.

COOKING

One of the easiest methods of cooking onions is merely to throw them into a baking tray unpeeled

ABOVE The author's favourite version of onion tart, made with uncooked onions (see overleaf)

and let them roast in a medium oven (190°C/375°F/gas5) for about 1½ hours (the size will dictate the exact length of time). Then take the outside skin off and the inside will be moist, sweet and cooked through. If barbecuing, do the same: cook them unpeeled and, as the duration of time will be shorter when the fire is hot, choose small onions so that they are thoroughly cooked.

The eighteenth-century English cookery writer Hannah Glasse gives many recipes using onions, including one for *ONION PIE*. In this, onions, potatoes, apples and eggs are all layered with bits of butter and seasoning strewn between them, six spoonfuls of water are added, then a crust is placed over the top and it is baked for 1½ hours. It is not clear whether the eggs are just beaten or have been hard-boiled, but I suspect the latter as she suggests adding 12 eggs to 450 g / 1 lb each of onions, apples and potatoes. She adds mace, nutmeg, salt and pepper as her seasonings. I have made a pie inspired by this recipe which omits the eggs and changes the seasoning to sage, marjoram and parsley. It is a good autumn or winter pie and makes an excellent supper or luncheon dish with a green salad. Apple, onion and potato are highly satisfactory together.

One of the great classic dishes is *ONION TART* – *tarte à l'oignon* – which is a famous Alsatian dish. I confess I have eaten disastrous versions of it, where the onion has merged into a thick egg custard and tinged it all with an unbearable sweetness. I have also cooked tarts which have been unmitigated failures. I do not want my tart filling to be either bland or sweet, nor do I particularly want it to be rich in butter, egg yolk and cream, though the tart might benefit with just one of those ingredients.

Here then are some recipes that strive towards keeping the onions' fieriness. Make sure the pastry base is thin – a shortcrust pastry rich in butter is the best, though a puff (if rolled very thin and if made with butter) could easily be substituted. A pastry made with half wholemeal flour blends well with the onion flavour. For a 27 cm / 10¾ in tart tin with a depth of 1 cm / ½ in you will need just over 450 g / 1 lb of onions (ie, about 4 or 5 medium-sized onions), sliced across (thickly or thinly, it hardly matters). Sauté the onions very gently in 85 g / 3 oz of butter in a covered pan for about 30 – 40 minutes. Raise the heat and take the lid away to allow any liquid to evaporate and to caramelize the onions slightly – this takes about 2 – 3 minutes. Then let the onions cool while lining the tart tin with the pastry. Preheat the oven to 220°C/425°F/gas7.

Variations

- If you want the classic onion tart, you now mix in 2 egg yolks and 2 tablespoons of thick cream, a pinch of salt and nutmeg and plenty of black pepper. Then the tart is baked for 20 minutes.
- If you want to taste more onion in your tart, simply spread the pastry base with your onion mixture and bake for 15 minutes until the pastry is cooked through and the onion has caramelized more.
- A sharper, denser flavour can be achieved by adding 2 tablespoons of soy sauce to the onion mixture before raising the heat. Then, when the liquid has evaporated, add a tablespoon of white flour to thicken the onions. Season with salt and pepper, then spread this over the pastry and sprinkle 55 g / 2 oz of Gruyère and 30 g / 1 oz of Parmesan cheese over the top. Bake for 20 minutes.
- Before cooking, add 1 deseeded and chopped red chilli to the sliced onions. Then, before raising the heat, stir a tablespoon of sweet paprika into the mixture. Season with salt and pepper, then spread this over the uncooked pastry and bake for 20 minutes.
- To make my favourite onion tart do not cook the onions before, slice a little less onion (3 onions rather than 4) and toss them in a bowl with a tablespoon of good Dijon mustard and 1½ tablespoons of very good olive oil, salt and freshly ground black pepper. Pile these sliced onions on top of the uncooked pastry and bake in the oven for about 15 minutes, or until about one-third of the chopped onion has browned slightly. You should get not only the crunch of the almost raw, but some onions which have softened in their own steam and some which have caramelized.

This recipe does not need a tart shape for its pastry base, as there is no custard or sauce for the sides to hold in. So a flat round pastry shape, like a pizza, spread on a baking tin – with the onions piled up on it – is perfectly satisfactory.

French Onion Soup is another classic dish which, when cooked perfectly, justifies its standing as one of the great soups of the world. Like the onions for the tart when cooked in the classic manner, they need long slow gentle cooking so that they caramelize. In my opinion, the best recipe for this, *gratinée lyonnaise*, appears in *French Regional Cooking* by Anne Willan, where it is enriched with egg yolks and port. There are other French onion soups, however, like *tourin* (sometimes '*tourain*', '*thourin*' or '*tourrin*') of the Périgord and Bordeaux areas. This is made with goose fat, garlic and onions, egg yolks, bread and red wine. In Quercy they sometimes add some preserved goose or shredded kohlrabi.

The soup below is an adaptation of a recipe from *The Lady's Companion*, 1753, a cookery book which Martha Washington owned. How the recipe got its name is a mystery, for onions do not figure large in royal kitchens.

THE KING'S SOUP

55 g / 2 oz butter
½ tsp mace
450 g / 1 lb onions, thinly sliced
1½ tsp sea salt
1.1 litres / 2 pt milk
(use soya milk if concerned about fat intake)
2 egg yolks
handful of chopped parsley
croutons, to serve (optional)

Melt the butter in a large saucepan. Throw in the mace, onions and salt. Let the onions sweat, stirring occasionally, for a few minutes.

Pour in the milk, bring to a simmer and stand over a low heat for 30–40 minutes.

Pour a little soup into a bowl, add the egg yolks and mix thoroughly. Then pour the egg mixture back into the soup. Stir until slightly thickened.

Add the parsley and serve with croutons, if using.

FRIED ONIONS

These are an unexpected treat as long as they are floured before frying. Slice the onions in fairly thick rings, then pop the rings into a bag with a little plain flour and shake it well. Fry in a light oil, like sunflower, until they are golden and crisp. Dust lightly with paprika to serve.

ONION AND CHILLI SALAD

1 tsp garam masala
juice of 1 lemon
150 ml / ¼ pt yogurt
3 large onions, sliced
1 cucumber, chopped small
1 red chilli, deseeded and finely chopped
1 green pepper, deseeded and finely chopped
sea salt

Mix the garam masala, lemon juice and yogurt together with some salt. Place all the vegetables in a bowl, pour the yogurt mixture over and toss well.

Leave for about 20 minutes and then serve.

MOROCCAN ONIONS

3 tbsp olive oil
3 large onions, sliced
2 tbsp tomato paste
3 tbsp mixed sultanas and raisins
salt and freshly ground black pepper

Heat the olive oil in a heavy-based pan. Throw in the onions, sprinkle a little salt over them and let them cook over a low heat, stirring occasionally.

When they are softened, add the tomato paste and the dried fruit. Continue to cook gently, but stirring so that it does not burn. Cook for another 5 minutes, then adjust the seasoning. Serve hot or cold.

GRATIN OF ONION AND POTATO

450 g / 1 lb potatoes, thinly sliced
450 g / 1 lb onions, thinly sliced
30 g / 1 oz butter
55 g / 2 oz Gruyère cheese, grated
55 g / 2 oz Parmesan cheese, grated
sea salt and freshly ground black pepper

Preheat the oven to 180°C/350°F/gas4. Grease a shallow oven dish lightly with a little of the butter.

Arrange layers of sliced potato and onions in the dish, dotting each layer with a little butter, some seasoning and a sprinkle of a tiny amount of each cheese. Finish with a thicker layer of the mixed cheeses.

Bake for 40–60 minutes, until the top is bubbly and golden.

ONION POTATO PURÉE

This is a way of serving mashed potatoes which is unexpectedly good. The onion is left raw and added chopped small to the potato purée after milk and butter has been mixed in. The crunch of the onion adds a wonderful texture as well as flavour. This dish is excellent as an accompaniment to game, and if there is any over it can be made into potato cakes, dusted with breadcrumbs and fried in oil.

GARLIC

Allium sativum

This powerful little bulb is one of the most ancient of flavourings. It was one of the staple foods of the poor, the majority of the world's population, for it comprised the basic relish to be eaten with bread. Literary references to garlic in the ancient world are legion. Its beginnings in prehistory are thought to be in the area of the plains of western Tartary and from there it was transported over the whole of Asia (except Japan), North Africa and Europe.

The Egyptian garlic which helped sustain the labourers as they built the Great Pyramid had rather small cloves (around 45 to a head) and was of a light purplish hue, a colour we can still grow. Some garlic heads have been found in tombs, clay models of garlic bulbs have also been found in Pre-dynastic burials at Naqada (3200 BC). In the *Ebers Codex*, a medical work compiled around 1550 BC there are over 800 therapeutic formulae, 22 of them mention garlic for various ailments from headaches to childbirth.

Theophrastus praises Cypriot garlic to be used as a dressing for salads, he directs you to pound the garlic well for 'it increases wondrously in bulk making a foaming dressing.' Garlic also had magical properties for the Greeks and was taken before battle to make warriors strong. Garlic heads were also placed on piles of stones at crossroads as a supper for Hecate, the goddess of enchantment and spells. Indeed, garlic from antiquity gathered its reputation to ward off evil. Greek midwives crushed the cloves in the room in which the child would be born and a garlic clove might well be fastened about the baby's neck.

Hippocrates (460 BC) praised the therapeutic qualities of garlic – good for wounds and toothache among many other ailments – as did Aristotle, a hundred years later. Dioscorides (AD 40) was a Greek physician who travelled as a surgeon with the Roman armies, giving him the opportunity to study the distribution and qualities of a host of plants. He noted the effects of sleeping potions made from opium and mandragora, which is used for surgical anaesthetics. His observations on garlic are particularly acute: after noticing the various types – the white, the purple and the wild (a type which has escaped from cultivation) – he goes on to write about its efficacy: 'it eliminates tapeworm, with wine it is good against snake bite, soothes coughing, clears the arteries, heals eczema and much else'. He concludes by adding that a purée of crushed garlic and black olives is a diuretic.

As with onions, garlic was disapproved of by the élite of both Greece and Rome. It was common people who ate it and, therefore, smelt of it. In fact, it is surprising that those parts of the world that were still not conquered by Rome could not smell the advance of the Roman legions, who ate great quantities of it. Not only was garlic thought to be an aphrodisiac, but as with the Greeks it was believed to make men strong and powerful. Pliny includes sixty-one garlic remedies from his *Natural History*, to take it as an aphrodisiac it should be 'crushed and beaten up with fresh coriander and taken in pure wine.'

But what of garlic in cooking? French gastronome Marcel Boulestin placed his finger on the pulse when he declared: 'It is not an exaggeration to say that peace and happiness begin, geographically, where garlic is used in cooking.' Where would the cuisines of France, Italy, Spain, Greece, the Middle East, the Slavic and Balkan countries and much of the Orient be without it? Surely it is the most powerful ingredient in the history of cooking, used as we have seen from earliest times. The Chinese text *Shih Chins*, or *The Book of Songs* (600 BC), describes the life of the warrior farmers on the north-western highlands of the Shensi province. At a spring initiation rite, held after the fields had been ploughed, they sacrifice a lamb seasoned with garlic and comment, 'What smell is this, so strong and good?' Garlic is mentioned in a verse of Ibn al-Mitazz, the tenth-century Caliph and poet, 'here pungent garlic meets the eager sight / And whets with savour sharp the appetite...'

The garlic sauce, forms of *aïoli*, whether emulsified with egg and oil or not, and garlic soup permeate all countries and cultures.

NUTRITION AND MEDICINAL

Garlic contains calcium, phosphorus, iron and potassium, also thiamine, riboflavin, niacin and vitamin C.

Its two volatile oils, allin and allinase are separate inside the bulb, but once cut and crushed the two mix and become allicin; this is the sulphur compound of the pungent odour we know so well. It is also a powerful antibiotic: seventy-two separate infection agents can be deterred by garlic. Cooking will destroy this antibacterial agent, though it leaves untouched other therapeutic qualities (see page 20).

Since 1983 over 130 scientific papers have been published giving evidence of garlic's powers to retard heart disease, strokes, cancers and a wide range of infections. It lowers blood cholesterol levels astonishingly quickly and increases the assimilation of vitamins. Blood tests have revealed that the most ardent garlic eaters display the least clot factor, an important element in helping to prevent heart disease.

Raw garlic kills bacteria, boosts immune functioning and probably helps to prevent cancer. One or two cloves a day is all you need. I take mine in the vinaigrette dressing with the daily salad: nothing is more delicious or more trouble-free. I gladly confess to enjoying raw garlic in sandwiches too, seasoning avocado and lettuce or tomato and mozzarella.

Cooked garlic can lower blood cholesterol, helps keep blood thin, acts as a decongestant and cough medicine and helps prevent bronchitis. Lastly, there is no evidence which suggests that the garlic preparations sold by health-food shops are in any way equal to the real thing, So stick to the use of garlic daily in your cooking. For the last thirty years I must have consumed about three or four heads a week.

GROWING, CHOOSING AND STORING

Garlic is very easy to grow even in a climate like Britain's. The individual cloves have to be sown about 15 cm / 6 in apart at the end of October or in November. Choose cloves which have come from more northern climes and are hardened to cooler springs. They will sprout quite soon and stay throughout the winter, through snow and frost. By March you will have healthy young garlic and you can use the green tops in a host of dishes. In Spain and southern France you will find sheaves of young garlic in the marketplace at this time. The main crop is ready in June or July.

Whenever possible try to use fresh garlic, though the outside may look dried the cloves will feel hard and be full of juice. Really fresh garlic is available in the spring and still has its growing stem which shows a little green. It has an excellent fiery flavour and some of the insides of the green stem can be used too.

How you buy your garlic depends on how frequently you use it. I can have strings of garlic hanging in the kitchen and they are not there long enough ever to dry out. If you use garlic modestly, however, buy a few heads and keep them in the salad drawer in the refrigerator or somewhere cool and shady.

Feel heads before purchase: if soft, they are old. Never buy heads which have started to sprout. Buy only those which are firm and freshly coloured or white; if a little dingy they have been around too long.

PREPARING

The easiest method of taking the outside paper skin off the garlic cloves is to blanch them, as one does tomatoes, in a little boiling water. The skin then slips off easily. If you crush garlic cloves with the help of the flat of a knife blade and the weight of your hand, it is not too difficult to extract the papery covering.

Cookery books used always to tell you to crush garlic with a little salt. How did this direction come about? For it is quite unnecessary; all salt does is to activate moisture so that the garlic will ooze its own juices more readily. If this happens on a chopping board it can be a nuisance. It was American food writer Richard Olney who taught his students never to use a garlic crusher, as it alters the flavour. I have never noticed this to be true and use one all the time.

Do not worry about garlic that is shooting, use what you can and – if you do not mind a great intensity of garlic flavour – use the green shoots chopped up in a salad. There is a myth that shooting garlic tastes bitter, but it merely tastes powerfully garlicky.

COOKING

If you wish to roast garlic you can simply throw some cloves in a baking dish with a little oil and leave them for 30 minutes in an oven preheated to 190°C/ 375°F/gas5, then eat the soft melting interior by cutting into the clove with a knife. There is an excellent recipe by Richard Olney in which you cook new pota-

ABOVE Garlic hanging to dry

toes with olive oil and plenty of unpeeled garlic in a covered pan over a low flame for 40–45 minutes. The garlic is partly fried, partly steamed and, once released by your knife and fork from its paper parcel, makes the most delicious purée for the new potatoes.

If you are as passionate about garlic as I am, then you will not mind peeling 40–50 cloves to be baked beneath a pheasant. I have quite happily peeled 200 cloves for garlic soup and I do believe that the more garlic you use in a dish the less pungent it ends up being. Cooked garlic in soup, for example, has a quite different flavour; it is both sweet and savoury, has a depth of pungency that is immensely satisfying and is warming to the spirit in the long winter months.

A splendid *GARLIC BUTTER* for toast as an appetizer or as a sauce for pasta can be made quite quickly by adding 10 or 12 peeled cloves to a little boiling salted water and letting them simmer for 5 minutes until soft. Then drain and mash the garlic with 85–115 g / 3–4 oz of butter and use for spreading. An alternative is to use olive oil instead of butter and blend the oil and the cooked garlic, adding one anchovy fillet. Again, you will find a garlic sauce made from poached garlic has a deeper and more fulfilling flavour than the more fiery variety made with raw garlic.

A similar *GARLIC SAUCE* can be made using, say, 20 peeled garlic cloves and poaching them in 575 ml / 1 pt of milk for 10 minutes, or until they are soft. Blend the milk and garlic to a smooth consistency and add 2 egg yolks to thicken the sauce. This is excellent with roast goose stuffed with apple and potato.

I am unrepentant in my love of garlic – the smell I inhale like perfume, the aroma on others' breath is a sign of life and its celebration. What angers me are directions for merely rubbing a garlic clove around a salad bowl – as if you are going to eat the utensil – and the snobbery which still links garlic odours with impolite society. In our age of deodorants and perfumes we tend to fly away from our own reality and this limits our awareness of the true nature of life.

AUBERGINES WITH PARSLEY AND GARLIC

for 6

This appeared in Elizabeth David's *French Provincial Cooking*. On rereading this classic book, which helped to revolutionize British cooking, it is astonishing to see how little garlic Mrs David used or recommended. I have thus changed the amounts.

2 large aubergines
olive oil
7 or 8 garlic cloves, chopped
handful of parsley, chopped
1 lemon
sea salt and freshly ground black pepper

Slice the aubergines thinly across in rounds. Place in a colander and sprinkle with salt. Leave for an hour so that they lose some of their juice. Rinse under a tap, otherwise they will be too salt, and pat dry.

Heat the oil in a pan and fry the aubergine slices. Take your time (Elizabeth David says 'fry them slowly'); they should be golden brown and soft. When done, keep warm and quickly fry the garlic.

At the last moment, add the parsley and whisk it around the pan. Pour this over the aubergine, then squeeze lemon juice over everything and season.

GARLIC SOUP

for 6

There are many variations which are all amazingly delicious. This is the simple version to begin with. What gives the soup its velvety texture is the emulsion of the puréed garlic in the olive oil. The flavour will not be much altered by the actual amount of garlic used, but the texture will. Hence, the more garlic, the smoother and thicker the velvet.

3–4 tbsp olive oil
3 heads of garlic, their cloves peeled
pinch of saffron strands
1.75 litres / 3 pt vegetable stock
(2 good-quality stock cubes)
sea salt and freshly ground black pepper
3–4 tbsp finely chopped parsley or chives

Heat the olive oil in a saucepan. Throw in the garlic and saffron. Lower the heat, put a lid on the pan and let the garlic sweat for 3–4 minutes.

Pour in the stock, bring to the boil and simmer for 15 minutes. Let cool, then liquidize to a thin purée.

Reheat, adjust the seasoning and add the parsley or chives before serving.

VARIATIONS

- Add two egg yolks (as described for the King's Soup, page 22) to thicken and enrich the soup.
- Add garlic croutons with the soup on serving.
- Add 85 g / 3 oz ground almonds with the stock.
- Add 225 g / 8 oz mashed potato before liquidizing.
- Add several softened bread crusts before liquidizing.

SKORDALIA

There are almost as many versions of this garlic sauce or appetizer as there are Greek islands, for almost every island has their own method of making it. The basic ingredient after the garlic can be potato, almonds, walnuts or bread; or you can have a mixture of garlic, nuts and bread. It is eaten with fish or on its own as a mezze or part of a first course.

Greek restaurants here often feature it, but I have never tasted a good one. So it is worth making.

1 head of garlic, cloves peeled
3 or 4 thick slices of wholemeal bread, soaked in water then well squeezed
juice from 1 lemon
55 g / 2 oz ground almonds
4 tbsp olive oil
sea salt and freshly ground black pepper

Place the garlic cloves in a blender jar and crush them to a pulp. Add the bread, lemon juice and almonds, season generously and blend again.

Then add the olive oil slowly until the mixture is the consistency of treacle – thick but a little runny.

RIGHT *Garlic soup and roast garlic*

LEEK

Allium porrum

This third vegetable of the onion family also goes back far into antiquity, gathering an enthusiastic following. Our modern leek has a resemblance to the wild leek (*Allium ampeloprasum*) which is a native of the Mediterranean region and the Atlantic islands of the Azores, Canaries, Cape Verde Islands and Madeira. It is this leek which was found in the remains of Jericho as early as 7000 BC. Other names for the wild leek are 'great-headed garlic' and 'Levant garlic'. It is a hardy perennial, remarkable for the size of its bulbs. For thousands of years it has been eaten raw by the peoples of the Mediterranean and the Middle East.

The earliest written records of leek cultivation are Egyptian and date from 3200 BC, 'fix his allowance at a thousand loaves of bread, a hundred jars of beer, one ox and a hundred bunches of leeks.' It was thought then and later that the leek originated in the Middle East. As we have noticed, however, this area was the first to domesticate plants and the leek was one of the favoured vegetables of summer. Pliny thought that the best leeks came from Egypt and also Aricia in Italy. They had a great vogue in Rome, for the emperor Nero ate them for several days once a month in the belief that they cleared and strengthened his voice, so that he could sing and declaim his works – thus earning the nickname 'porrophagus'.

Leeks remained popular throughout the Dark Ages. Charlemagne's list of foods to be cultivated in his domains includes them, and their success did not decline in the Middle Ages. The classic herbal of the fourteenth century (*The Four Seasons of the House of Cerruti*) tells its readers to choose pungent-smelling leeks to be cooked and mixed with honey and swallowed slowly. Not only an aphrodisiac, it declares, but also a diuretic which helps chest problems, gets rid of catarrh and cleanses the lungs. Nero's example is obviously remembered, if not mentioned.

Leeks have become the national emblem of Wales, recalled by Shakespeare in *Henry V* when Fluellen wears his leek in his hat. Pistol announces 'tell him I'll knock his leek about his pate upon Saint Davy's day.' The origin of wearing a leek on St David's Day on 1 March is in memory of an ancient victory where the Welsh identified themselves with the vegetable, wearing it like a feather in the cap.

Thomas Tusser, farmer and poet, wrote in 1557: 'Now leeks are in season, for pottage full good, / And spareth the milch-cow, and purgeth the blood, / These having with peason, for pottage in Lent, / Thou spareth both oatmeal and bread to be spent.'

People feel slightly equivocal about leeks. On the one hand they have been praised as being a great delicacy, having a far subtler flavour than the rest of the family; useful as a purée and mixed with egg yolks in a classic *flamiche aux poireaux* or the leek tart of Florence, *porrata*, the pastry of which is made with yeast and eggs. Or, people disparage leeks as being tainted by the close relationship with garlic. They are, in fact, 'a poor man's asparagus'.

Jane Grigson notes that leeks fell from grace for all of three centuries. They began to disappear in the sixteenth century and they were not much applauded when they crept back in the nineteenth. Mrs Beeton's classic nineteenth-century work gives only two recipes and admonishes the cook to boil the leeks well or they will 'taint the breath'. So I fear here we are back again on the basic fear of human stench.

NUTRITION

Nutritionally, leeks have much of the properties of onion and garlic, are even richer in nutrients and have a higher protein content; hence they are good to eat as a staple ingredient in everyday cooking.

VARIETIES

Leeks are useful vegetables as they can be grown throughout the winter months, from September to April. The early varieties are Marble Pillar (it has a long stem) and Early Market, which are both tasty and mild. Musselburgh, a Scottish variety with slim stems, Argenta and King Richard are harvested in midwinter. For the spring, Giant Winter Catalina and Winter Crop (the hardiest variety of all) are available. There is little difference in flavour among the varieties, but some have darker green leaves than others and are excellent for colouring soups and purées.

CHOOSING AND PREPARING

Choose firm leeks where the green ends are not discoloured or fading. Small leeks are useful for some dishes where they will be served whole; but for soups, purées and tarts, where the leeks will be sliced, large leeks are less fiddly to prepare. Clean them by cutting off the coarser green leaves, then splitting the leek down the middle so that any soil and grit can be washed away. If you wish to keep the leeks whole, then simply split through the green to where the dirt is and soak or rinse. Be careful when rinsing leeks not to drive grit deeper into the vegetable, so it is wiser to hold them green ends down.

COOKING

For soups *et al*, just slice diagonally as thinly as you want, across the leek and place in a pan with a tiny amount of olive oil, butter, or both, and place over a low flame. Leave to steam in their own moisture for 5 minutes or until soft, until the next stage.

Dorothy Hartley in her seminal *Food in England* (1954) claims leeks braise better than they boil and suggests you cook them whole in a closed well-buttered dish with the addition of a little milk or stock. 'It is a mistake,' she says, 'to serve these delicate vegetables with thick sauce; the liquor and butter together in the dish should be sufficient.'

What would she have made, I wonder, of this Roman recipe? Chop 450 g / 1 lb of leeks and sauté in 2 tablespoons of olive oil and 2 of white wine, add 55 g / 2 oz of chopped green olives and 1 tablespoon of *nam pla* (Indonesian fish sauce), which is our substitute for the Roman *liquamen* (see page 143). Simmer for 5 minutes and serve.

Here is another excellent medieval leek recipe which also uses wine. Wash and trim the leeks, then slice in half lengthwise. Cook in 2 tablespoons of olive oil and a glass of white wine and season. The

BELOW *Leeks on their way to market*

leeks will cook in about 5 minutes. Have some fresh crusty toast ready and pour the leeks with the sauce over it.

Italian food writer Anna del Conte tell us that when leeks are very young they can be served raw, but she suggests that they are first soaked in cold water to get rid of their strong flavour, then cut into thin rounds and dressed with olive oil, vinegar, a little mustard, salt and pepper. I use them raw, without the soaking, cut paper-thin and often added to other raw winter vegetables – such as grated carrot, beetroot or celeriac. Particularly the last, as they add a great deal of zip to a celeriac *rémoulade*. (See page 221). Anna also gives a marvellous leek dish from northern Italy which she says is an old favourite, *Porri alla Milanese*. This would make an excellent luncheon or supper dish for two. She says this is the poor man's version of *asparagi alla milanese*. I have changed the method slightly: in the original version the leeks are boiled first.

Wash and trim 6 leeks, slice them in half lengthwise and place them in a buttered baking dish. Dot the leeks with a little more butter, season and cover the dish. Place in an oven preheated to 190°C/375°F/gas5 for 20 minutes. Uncover the dish and sprinkle 55 g / 2 oz of Parmesan cheese over the leeks. Leave for another 5 minutes while you fry 4 eggs in a little more butter, being careful to cook the white but leave the yolks runny. Then slide the eggs over the leeks. Eat with good crusty white bread to soak up all the delicious juices.

The Welsh dish *cawl* (pronounced cowl) is a national leek dish. It is made with meat and any amount of vegetables, though leeks must always feature. It lies somewhere between a soup or broth and a stew. In fact it can be eaten in two stages, the broth eaten first and the meat and vegetables second. In Wales they also have many variations of leek soup; it is indeed almost exactly the same as the French *potage à la bonne femme*, or if cooled and with some added cream – *vichyssoise*. Leek soup, I believe, is one of the great soups of the world. It can hardly fail: the fusion of that stringent aspect which all the onion family have with the bland earthiness of the potato is always immensely satisfying.

SIMPLE LEEK AND POTATO SOUP

450 g / 1 lb leeks
450 g / 1 lb potatoes
30 g / 1 oz butter
1.75 litres / 3 pt vegetable stock (see note below)
sea salt and freshly ground black pepper

Chop the leeks across into chunks and peel and dice the potatoes.

Melt the butter in a pan and throw in the vegetables. Let them sweat for a minute or two, stirring with a wooden spoon.

Pour in the stock and simmer for 15 minutes. Leave to cool, then blend. Return to the pan and reheat. Taste and adjust the seasoning.

Note: the success of this soup hinges on the stock you use. It must not be too pervasive. The leek and the potato must be the most prominent flavours, yet though you can use just water, I believe it benefits from a slight sub-text of savouriness from the stock. There are now a variety of excellent vegetable stock cubes, experiment and find one you can rely on.

Variations

- For a more interesting texture, blanch half the soup and return it to the pan.
- Add 300 ml / ½ pt of single cream after blending and reheat carefully without letting the soup boil.
- Add a spoonful of sour cream and croutons or a sprinkling of Parmesan when serving the soup.
- Add a glass of dry sherry before serving.

In the spring and summer months the soup, in the *vichyssoise* tradition, can be served chilled.

STIR-FRIED LEEKS

I use this method of cooking leeks more than any other, because I like to eat leeks when they are young and fairly slender.

450 g / 1 lb youg leeks, trimmed and cleaned but left whole
30 g / 1 oz ginger root, peeled and grated
1 red chilli, deseeded and chopped
3 garlic cloves, chopped

1 tbsp sesame oil
pinch of sea salt
pinch of sugar
few drops of light soy or dry sherry (optional)

Chop the leeks into 7.5 cm / 3 in lengths.

Heat the sesame oil in a wok or a frying pan. Throw in the ginger root, chilli and garlic, followed by the pieces of leek. Stir-fry vigorously for about 3 minutes, or until the leeks have softened on the outside but are still *al dente* inside.

Add the salt and sugar at the end and also, if you wish, at the last minute you can add a few drops of light soy or dry sherry.

GRATIN OF LEEK AND POTATO

I like the flavour of leeks and green peppercorns together. I have used this mixture in a tart and a roulade, but one of the simplest of dishes consists of leeks served with a white sauce studded with green peppercorns. The following recipe, however, is another variation – a gratin.

675 g / 1½ lb leeks
450 g / 1 lb potatoes
2 tbsp dried green peppercorns or ones in brine which have been carefully drained
55 g / 2 oz Gruyère cheese, grated
55 g / 2 oz Parmesan cheese
575 ml / 1 pt single cream
sea salt

Preheat the oven to 150°C/300°F/gas2. Chop the leeks diagonally and slice the potatoes thinly.

In a shallow earthenware or oven dish, arrange layer of the vegetables, starting with the leeks followed by the potatoes. Sprinkle each layer with a few green peppercorns and a little of each cheese and some salt. Continue the layers until all the vegetables are used up, finishing with the potato.

Pour the cream over the dish and conclude with a dusting of cheese. Bake in the preheated oven for 2 hours. The slow cooking allows the potatoes to soak up the cream and flavouring.

SHALLOT

Allium ascalonicum

Named after the ancient Palestine port of Ascalon (now modern Ashqueion, in south-west Israel) where it was thought to originate, shallots differ from the common onion in that they bunch up with bulbs that multiply freely. They make a separate group with the 'potato onion' or 'the multiplier onion'. Both Theophrastus and Pliny mention the Ascalon onion with a degree of admiration, noting how good it was when used in sauces.

John Mortimer, the agricultural writer in *The Whole Art Of Husbandry* (1707) says of shallots, 'They give a fine relish to most sauces and the breath of those that eat them is not offensive to others...' Mortimer is quite right. Shallots have a delicate flavour and a less stringent smell than most other members of the onion family. They also have the capacity of dissolving easily into a liquid when they are cooked, hence they are of an immeasurable use in the creation of sauces.

They are a vital component of many of the most noted classic sauces of French haute cuisine: eg Bercy, béarnaise, bordelaise, chivry and duxelles. They are also a classic accompaniment to grilled or fried steak or cutlets and are best, in my view, if added to a little red wine, mustard and salt and then barely cooked at all so that the chopped pieces of shallot still have plenty of bite and texture to them.

VARIETIES

One of the most common of shallot varieties is Dutch Brown, but there are many more, including Golden Gourmet, Sante, Sante Red and Atlantic.

Shallots come in white, red and yellow varieties. Their shape can vary from that of a small pear to elongated types that, once peeled, are easily sliced across thinly for use in salads – I am thinking of a type called Giant Long Keeping Red which also has a yellow sibling.

SWEET AND SOUR SHALLOTS

for 6

I am fond of shallots, eaten either hot or cold as in the following recipe which makes an excellent first course on a bed of salad leaves.

450 g / 1 lb shallots
30 g / 1 oz butter
2 tbsp red wine vinegar
1 tbsp muscovado sugar
pinch of sea salt

Peel the shallots and throw them into some salted boiling water. Simmer for 3 – 4 minutes, until they are just soft on the outside. Drain.

To the same pan, add the remaining ingredients. Stir until the sugar has dissolved and the shallots are glossy. Leave to cool.

These are delicious warm or cold. To serve, simply pour a little sauce and a few of the shallots over some salad leaves arranged on individual plates.

CHIVE

Allium schoenoprasum

These green tufts of aromatic grass are perennial and indigenous to the whole of the northern hemisphere from Arctic Europe to Japan. They may be found from the Mediterranean shores to Sweden and across North America. The bulbs are tiny and it is the spear-like leaf which has always been used for flavouring. The American botanist Sturtevant noted in the nineteenth century that chives were much used in Scotch families and are considered 'next to indispensable in omelettes'. The English herbalist Gerard thought they 'made a pleasant sauce and a good pot-herb'.

Certainly we do not make sauces using chives, as the fugitive flavour does not really survive lengthy cooking, but it appears with unceasing regularity in herb mixtures and is particularly useful mixed with soft cheeses. Unfortunately, the commercial cheeses might look attractive speckled in green, but the delicate chive flavour rarely comes through. It does, however, when you use chives in home cooking, such as omelettes and roulades, or mixed with mashed potato and in vinaigrette when used as a dressing over fish or vegetable moulds. Sometimes, in fact, the whole leaf can be used like a green spear, so that it appears like three green stripes in presenting a dish.

Lately, too, the chive flower has been pressed into service. This is also edible, but few people enjoy consuming flowers as much as I do. Try picking the flower buds and using them in a salad.

WELSH ONION

Allium fistulosum

You will not usually find Welsh onions in shops, but they are quite popular with gardeners. Other names are ciboule or Oriental or Japanese bunching or everlasting onion. In fact, they have nothing at all to do with Wales and probably originated in Eastern Asia or Siberia. This is the favourite onion of China and Japan, and is perfect for flavouring as the green leaves can be cut off and the bulb then simply grows more.

For centuries this onion has been a staple in the cuisine of the East. Joy Larkcom calls it 'the most ubiquitous of Chinese vegetables. In Taiwan I drove through a famous onion village, where long onions were hanging to dry from every window and roof.' The advantage of the bunching onion is that different parts of it can be harvested at different times – the small leafy shoots first for flavouring and salads, the small scallion or spring onion next and then the long mature blanched stem last. The Latin *fistulosum* refers to the hollow stems.

ROCAMBOLE

Allium scorodoprasum

Rocambole was once thought to be a form of garlic and it certainly has the flavour, though on a more delicate scale. It is also called sand leek and Spanish garlic. It grows wild in southern Europe and on the Mediterranean coast, especially the Greek Islands. It can be identified by a stem which coils in the upper part and by the fact that its seed head is a cluster of bulblets.

There is no record of the plant ever being cultivated in antiquity, though the herbalist Gerard mentions it as a cultivated plant in 1597. The seventeenth-century English diarist John Evelyn prefers it to garlic, for that 'is not for ladies palettes, nor those that court them, rather...a light touch on the dish with a clove thereof, much better supplied by the gentler Roccombo.' By 1718 Richard Bradley, the botanist, laments 'The Rocambole, for its high Relish in Sauces, has been greatly esteem'd formerly, but now a-days is hardly to he met with', adding that 'considering how small a Quantity of it is sufficient to give us that Relish which many onions can hardly give, it ought to be preferred'. Only the bulblets in the head of the rocambole were added to the sauce.

Soon after, rocambole disappears from our kitchen gardens completely. Now, it is virtually unknown, unless you find it in some unspoilt part of the Mediterranean.

ROCAMBOLE *Allium scorodoprasum*

ABOVE Chinese chives

CHINESE CHIVE

Allium tuberosum

These have been used for centuries throughout the East. They can be grown here, but take time to get established. They are worth the effort as they have a most attractive but delicate garlic flavour. Use them like chives. All the parts can be used, both the white and green of the stalk and the bulbs.

They are also a decorative asset to the garden, as they remain green throughout the spring and summer without the browning that our own chives soon develop. Some supermarkets stock Chinese chives at times. You can identify them by the white bulblets at the tip of the leaves that look like flowers.

Liliaceae

ASPARAGUS

Asparagus officinalis

Asparagus is a member of the lily family (Liliaceae) which is allied to the onion family. In its wild state it grew around the seashores and the river banks of southern Europe and the Caucasus. It was the wild plant (*Asparagus aphyllus*) that was known and loved in ancient Greece, but the Romans cultivated it – though, following Cato's instructions, it would have been from the wild seed. The growing of asparagus was obviously an act of love. Pliny says, 'of all the plants of the garden it receives the most praiseworthy care'. Yet he adds, 'nature has made the asparagus wild, so that anyone may gather as found'. Then he tells us of a much-manured asparagus at Ravenna, three spears of which weighed a pound. The Roman pound weighed 0.721 of our pound, so I estimate that each spear would weigh about 100g / 3½oz which is very impressive but no bigger than some giant stems you might find now. The Latin poet Martial also praises Ravenna asparagus, saying it was the best in the world. Knowing the Romans, they would have been sure to have grown it for flavour as well as size.

For there is no doubt that the Romans loved the vegetable and also knew how best to cook it. The Emperor Augustus coined a phrase '*velocius quam asparagi conquantur*', meaning to do something faster than you can cook asparagus. Julius Caesar first ate it in Lombardy and wanted it served with melted butter. Though the ancient world recorded its love affair with this plant, it then strangely vanishes entirely from history. However, much later in the seventeenth century we have John Evelyn waxing enthusiastically over English asparagus being 'sweet and agreeable though of moderate size', better by far than the Dutch kind, which was large and great due to 'the rankness of the beds'.

Another wild species (*Asparagus prostratus*) grows wild around the grassy sea cliffs and dunes of northern Europe and was almost certainly eaten from the earliest times, the first moment in March when the edible shoots showed their tips above the surface. It is one of those plants which likes salt, hence it will feel at home close to the sea. You may well come across the Mediterranean wild variety in the markets of southern Spain and Italy; bunches of very spindly green stalks appear on stalls in the spring. The flavour is strong and rather bitter, which is why it is often boiled and chopped, then mixed with eggs. The flavour is then highly satisfactory, for the bland creaminess of the eggs offsets the salty tang of the wild plant.

There are three types of cultivated asparagus: white, purple and green. All three are grown and enjoyed in France. Other countries tend to have their particular favourite; for instance, Spain mostly cultivates the white, much of the harvest being canned. White asparagus never sees the light; the furrows in the fields are piled up (like potato cultivation) in rows and there is not a plant to be seen. I asked a Spanish farmer how on earth anyone finds the asparagus to harvest it. With great pleasure he showed me the hair cracks along the mound, evidence of that thrusting spear below. When the crack was seen, the knife went in, following the spear down and slicing it 15 cm / 6 inches below. For me, alas, the white has little flavour, nor do I find its blanched aspect has much appeal. The purple variety is allowed to grow 2.5 – 5 cm / 1 – 2 in above ground before it is cut and this has more flavour than the white. It is, however, the green which we go for in Britain, America, Italy and France and it is unquestionably the finest.

When the asparagus is thin and pencil-like it is called sprue and is frequently used in Italy, often with the addition of grated Parmesan – the flavour of which fuses well with that of the asparagus.

Because of its shape and its method of consumption, asparagus has always been considered an aphrodisiac. Though the Emperor Augustus would hardly have approved of the method of cooking recommended in an Arab love manual, where the asparagus is first boiled, then fried in fat and then covered in egg yolk. A daily portion of this dish, it was said, would keep the virile member alert night and day. Both Gerard and Culpeper wrote of the vegetable 'increasing seed and stirring up lust'. Louis XV's mistress Madame de Pompadour – with her eagerness for aphrodisiac foods – lived off asparagus tips and egg yolks with an occasional dish of truffles, celery leaves and vanilla. Her name is given to a recipe – *asperges à la Pompadour* – which uses egg yolks, but unfortunately it has cornmeal to thicken the sauce, an ingredient she was unlikely to have had at her disposal.

The most striking characteristic noticed by people who have consumed asparagus is the strong pong that emanates from the urine. It was noted by Dr Louis Lemery in his *Treatise Of All Sorts Of Foods* in 1702, 'asparagus causes a filthy and disagreeable smell in the urine as everybody knows'. The phenomenon is caused by the excretion of methyl mercaptan, triggered off by the sulphur-containing amino acid methionine in the asparagus. All who have eaten asparagus excrete the same. I stress this as there are some who claim they have never noticed their urine pong in such a way. The explanation is that our ability to detect this odour differs from person to person.

NUTRITION

Asparagus contains vitamins A, B2, C and E and calcium, copper, magnesium, iron and phosphorus. It is also an excellent diuretic and is also said to break up oxalic acid and crystals in the kidney; hence it is good at relieving rheumatism and arthritis.

VARIETIES

Until recently if you wanted to grow your own asparagus you had to buy crowns, plant them and wait for three years. Now, there are new varieties where you can begin cropping from the first year. It is, not surprisingly, the male plants that give the phallic spears and Accell produces all male seeds from the first year. Connover's Colossal is also in seed form and is ready for cropping in the second year. Both give you green stems.

CHOOSING AND STORING

Buy and enjoy asparagus in season, from the beginning of April to the middle of June in a good summer. The plant needs both plenty of rain and sun for a good growing season.

Fresh white asparagus can be bought in Spain,

while Spain also exports it canned. California exports asparagus throughout the winter, but I am a great enthusiast of eating foods in season. Canned green asparagus I would not bother about.

Fresh asparagus will keep well for some days (if you must, up to a week) in the salad drawer of the refrigerator. if you intend to do this, however, untie the bundle so that the spears are not pressed against each other and each is given room to breathe.

PREPARING

Most recipes direct you to peel the asparagus from the tips down to the woody stems. I cannot agree that this is at all necessary. The spears need only be rinsed briefly under a running tap, as everything is edible down to the fibrous end which may be cut off.

COOKING

Freshly picked asparagus is very good to eat raw. Evelyn notes that the spears were sometimes 'but very seldom eaten raw with oil and vinegar.'

What is the perfect way to cook asparagus? There is little doubt that an asparagus kettle is by far the most foolproof method. Drop the spears tips upwards into the wire basket. Bring a few inches of water to the boil in the kettle and drop the basket in. Place the lid on and leave for 5 minutes. The exact cooking time does depend upon the thickness of the spears and the massive spears of Ravenna would have needed more like 20 minutes. However, it is far better to have a spear with a little crunch to it.

If you haven't got a kettle, stand the bundle upright in a large saucepan and wedge it with new potatoes which you boil in the water. Cover the top with a dome of foil. No need to eat the potatoes at the same time, though their flavour is superb; they will make an excellent salad.

What to dip these spears into should not tax the connoisseur too hard. I agree with Julius Caesar, there is little doubt that melted butter is the most delicious, though I would add the juice of half a lemon to sharpen it. Both hollandaise and *aïoli* are excellent, but the simplicity of the dish is now blurred and the

RIGHT Green and white asparagus and sprue

calories are mounting up. A vinaigrette made from sherry or balsamic vinegar is as delicious as anything else and is perhaps the simplest to contrive.

However, sauces can also be made from sour cream flavoured with herb vinegar and nut oils, plus perhaps chopped herbs or spices. Greek yogurt or a low-fat plain yogurt with a few similar additions might be a slightly healthier alternative.

Quite another method of cooking which has lately found favour – ensuring an *al dente* spear with quite a bit of crunch – is grilling. Melt some butter in the grill pan and lay the spears in it, turning them so that they are well covered. Then grill for up to 5 minutes, turning the spears so that they are briefly seared on all sides. Alternatively, make up a mixture of olive oil and soy sauce and paint each spear with this.

QUICK ASPARAGUS RISOTTO

1 vegetable stock cube
450 g / 1 lb asparagus, trimmed and chopped
30 g / 1 oz butter
2 tbsp olive oil
3 shallots, sliced
6 spring onions, diced
115 g / 4 oz basmati rice
55 g / 2 oz Parmesan cheese, grated
sea salt and freshly ground black pepper

Bring about 250 ml / 8 fl oz of water to the boil, add a vegetable stock cube and throw in the chopped asparagus. Simmer for 3 minutes.

Meanwhile, in another pan, melt the butter in the oil. Throw in the shallots and spring onions, let them sweat for a moment, then add the rice. Stir so that the rice is coated with the oil and the onions are softened. Now add the asparagus and almost all its vegetable stock – about 150 – 175 ml / 5 – 6 fl oz. Stir the risotto, then cover and leave for 8 minutes.

Take a peep: if it looks too dry, add the rest of the stock; if it looks done, taste and see. Add a little seasoning and stir in half the Parmesan. Put the lid back on and leave for another minute.

Then turn out on a serving dish and sprinkle with the rest of the Parmesan.

ASPARAGUS SOUP

for 6

450 g / 1 lb asparagus
1 vegetable stock cube
300 ml / ½ pt double cream
sea salt and white pepper

Cut off the woody stems from the asparagus and discard. Cut off the tips of the asparagus and drop them into 1.5 litres / 2½ pints water with the stock cube. Bring to the boil and poach for 5 minutes. Take out the tips with a slotted spoon and reserve.

Now boil the asparagus stalks for up to 15 minutes, then allow to cool and liquidize. Taste and adjust the seasoning. Make sure there are no fibres; if so, put the soup through a sieve.

Mix in the double cream, add the asparagus tips and chill for an hour. Alternatively, you can reheat the soup to serve, taking care it does not boil.

ASPARAGUS SAUCE

This is good with egg or fish dishes or with crudités

12 asparagus spears
1 vegetable stock cube
1 tsp celery salt
150 ml / ¼ pt double cream

Cut off the woody stems from the stalks and discard. Then chop the stems into small dice.

Heat 450 ml / ¾ pint water with the stock cube and throw in the asparagus. Boil for 3 minutes. Leave to cool and then liquidize. Sieve to ensure there are no fibres. Mix in the celery salt and cream.

Note: This sauce can be turned into a mould by adding some gelatine or aspic jelly. Alternatively, add one beaten egg and place in a ramekin in an oven preheated to 190°C/375°F/gas5 for 20 minutes. Leave to cool and then unmould.

RIGHT *Grilled asparagus sprinkled with Parmesan cheese*

Chenopodiaceae

THE BEET FAMILY

SPINACH *Spinacea oleracea*
SWISS CHARD *Beta vulgaris* subsp. *cicla*
BEETROOT *Beta vulgaris* subsp. *vulgaris*
ORACHE *Atriplex hortensis*
GOOD KING HENRY *Chenopodium bonus-henricus*
MARSH SAMPHIRE or GLASSWORT
Salicornia europaea
QUINOA *Chenopodium quinoa*

Aizoaceae

NEW ZEALAND SPINACH *Tetragonia expansa*

This family includes, among other edible plants, spinach and beetroot. The tribe derives its botanical name, meaning 'goose foot', from the Greek 'chen' *(a goose) and* 'pous' *(a foot) – an allusion to the characteristic design of the leaves which is supposed to resemble the webbed feet of geese.*

They grow in temperate climes, often near the seashore, on salt marshes and waste ground. It is interesting to note that so many of our food plants grow near the seashore, underlying the fact that the earliest settlements were by estuaries and river delta – regions enriched by the minerals being continually washed down from the mountains, which living organisms then absorbed. The plants within the beet family contain large quantities of iron and many other minerals.

In all, spread across the world, there are about 600 members of the family. Quinoa was eaten in South America at the same time as spinach beet was eaten by the Romans.

Also in this section is New Zealand spinach, the one edible plant of the carpet-weed family (Aizoaceae) which includes ice-plants or sea figs and living stones. The former are Mesembryanthemum crystallinum, *a familiar sight in California, where the plant clothes many embankments with a thick fleshy green carpet dotted with purple daisy-like flowers.*

SPINACH

Spinacea oleracea

'Spinach is not worth much essentially,' writes the eighteenth-century French gourmet Grimod de la Reynière, adding, 'it is susceptible of receiving all imprints: it is the virgin wax of the kitchen.' One wonders what had occurred to Reynière's taste-buds that he could claim such nonsense. Spinach obviously is one of those flavours which has its fervent disciples as well as its detractors. For example, Louis XIV, forbidden spinach by his doctor, was supposed to have sent for it, saying, 'What! I am King of France and I cannot eat spinach?'

This vegetable, loathed by many young children, was made famous earlier this century as a health food by the cartoon character Popeye, who is given mythic strength through eating it. Alas, the strip cartoons and animated films never achieved their aim of boosting the consumption of the vegetable (or the consumption of green vegetables in general by Americans). I am, however, a fervent disciple and love it for its powerful flavour.

Spinach is not mentioned in antiquity, but Swiss chard – a very similar leafy vegetable – was noticed by Aristotle and many others, so it is odd that spinach was neglected. It is, after all, too individual and powerful in flavour to have been overlooked if it had been grown either in its wild or cultivated state. If it had been growing, the Greeks surely would have discovered it when Alexander conquered Persia, for it was certainly growing wild there and I suspect would have been used in Persian cooking with other wild leaves and herbs.

The first mention of spinach plants was 800 years later, as a present from the King of Nepal to the Chinese Emperor in AD 647. It was obviously a success because, by the ninth century, it is recorded in a work on Chinese agriculture and we know it was continually grown in the Imperial gardens and parks. It had come to Nepal from Persia where it was regularly cultivated. I imagine its cultivation there must have been recent, for the Greeks – and certainly the Romans – would have begun using it if the plant had been growing any earlier.

The name derives, via the Arabs, from the Persian word '*espenaj*'. From Persia it was accepted by the Arabs in their cooking and came to Europe through the Moorish occupation of Spain. Spinach is much richer in protein than other leaf vegetables, with a high vitamin A content. It was valued by Arab physicians and its first appearance in a cuisine must have been in those highly spiced dishes of the Middle East, where its strong mineral flavours would have stood up well to the addition of many spices – totally unlike the 'virgin wax' of Reynière's statement.

Albertus Magnus (1200 – 80), the teacher of St Thomas Aquinas and disciple of Aristotle, knew the prickly-seeded form of spinach, so the vegetable had moved north from Spain into Germany by that time. It came to be associated with Lent, as the seeds if sown in winter would leaf in time to be used for Lenten dishes. It appears in a list of vegetables recommended for monks on fast days in 1351.

Spinach appears as 'spynoches' in the earliest English cookery book, *The Forme of Cury* (compiled 1390), in which the most popular vegetables are cabbage, leeks and radishes. Catherine de Medici, the daughter of Lorenzo, who married Henri II of France in 1533, was said to be very fond of spinach, so much so that in French cooking the description *à la florentine* indicates that the dish contains spinach.

It is the prickly-seeded form that is the original plant. We now call it winter spinach, though with modern improvements both the prickly seeds and the round seeds can be sown in autumn for late winter harvesting. Smooth-seeded spinach made its first appearance in the sixteenth century; it is of a lighter green in colour and not quite so intense in flavour.

Spinach was popular as much for its colouring as its taste. Spinach water was used to colour cakes and desserts. It was also used to make touch-paper for fireworks, as paper that had been soaked in it and then dried out smoulders well.

By 1536 spinach had become thoroughly acceptable, known in both England and France. Elinor Fettiplace in her *Receipt Book* uses spinach as a sauce with chicken, either as a purée or cooked with parsley, mace, currants, raisins, dates, prunes and diluted with half a pint of sack. Murrel's *Two Bookes of Cookerie* (1638) gives a recipe for French Puffs with Green Herbs. Sweet spinach tarts had become popular.

The Tudors inherited the medieval love of colouring; the dark green spinach studded with dried fruits looked as splendid as it must have tasted. (It is interesting to see how ideas survive. Mrs Leyel, the English herbalist, cookery writer and founder of Culpeper House, in 1925 publishes a recipe for spinach purée studded with glacé cherries to be served on ice with cold tongue.)

By 1747 Hannah Glasse, in her *The Art of Cookery Made Plain and Simple* – the only really important English cookery book of that century – tells her readers how to cook spinach without water in a tin box tightly closed over a fire. She also gives how to dress spinach by first stewing it, then draining and chopping it and adding half a pint of cream, salt, pepper, grated nutmeg and a quarter pound of butter. This is simmered over the fire for a quarter of an hour, stirring often. A long French roll is sliced, fried in butter and any number of poached eggs are served with the spinach. Serve it for supper, Hannah Glasse suggests, or as a side dish at the second course.

Adam's Luxury and Eve's Cookery in 1764 still includes a recipe for a sweet spinach tart mixed with dried fruits and candied peel and two other sweet recipes, but creeping in (influenced no doubt by Hannah Glasse) is a savoury – a spinach purée with cream and pepper topped with poached eggs, which could well be served now.

Spinach had reached America by the 1800s. Thomas Jefferson grew it in his garden.

NUTRITIONAL AND MEDICINAL

Spinach and its relatives have a long history of medicinal use. The darker green the vegetable or leaf, the richer in minerals and vitamins it is; so spinach is one of the very best vegetables in nutritional terms. It contains plenty of vitamin C, A and B and some vitamin K as well as potassium, calcium, magnesium, iron, iodine and phosphorus.

Cooked spinach yields up oxalic acid, however, which forms insoluble compounds with the calcium and iron and therefore halts the absorption of these minerals by the body. So if you wish to gain all the nutrition from the dark green leaves, it is far better to eat them raw.

In some recent studies, spinach has thought to be a help as an antidote to lung cancer, because of its high concentration of carotenoids – including beta carotene. Raw spinach has more carotenoids (36 mg), than raw carrots (14 mg). It is also reputed to lower blood pressure.

VARIETIES

You can grow spinach throughout the year, so it is never completely unattainable, though the quality is not all it might be when the plant gets too old and the leaves too large, ragged and coarse. Spinach is always at its best when it is in its infancy and the leaves are small. It divides into summer and winter varieties, so the summer varieties can be picked small from late spring, while the winter varieties can be picked small from mid-autumn onwards.

Of the summer varieties, an old favourite is King of Denmark, which has round leaves growing high off the ground; another is Cleanleaf or Bloomsdale. Symphony and Norvak have large leaves and are resistant to bolting.

Nearly all the winter varieties have prickly seeds which means that they are frost-hardy. There is Broad Leaved Prickly and Long Standing Prickly. Monnopa

---

SPINACH *Spinacea oleracea*

has less oxalic acid content than most (see above), so is recommended if you have young children to feed.

CHOOSING AND PREPARING

Make sure to buy spinach leaves where the stalks are firm and the leaves springy; thus you will know that the leaves have been picked early that morning. Discard all the limp and discoloured leaves, really old spinach goes not only limp but yellow. You will need to buy about 450 g / 1 lb to get enough for two people. Store spinach in the salad drawer of the refrigerator and don't keep it for longer than about two days.

To prepare: cut off the stems and, if they are thick, tear the leaf part away and discard the central ribs. Wash the leaves well in cold running water, then drain and shake them dry.

All the old cookery books are keen on washing the spinach leaves with care several times. Industrial Britain was, of course, a lot dirtier than today and where market gardens were near big cities, as they often were, the leaf vegetables accumulated soot and grime on the leaves.

COOKING

There are two schools of thought on how to cook spinach: firstly with no water at all, secondly with masses. I am against the gallon of water approach and prefer to cook my leaves without any added water.

After washing and draining the leaves, place them in a thick saucepan with a little salt and leave over a low heat. The leaves simply sweat and steam, cooking in their own moisture. About 450 g / 1 lb of spinach should be cooked through within 5 minutes, but the exact time depends on the amount of leaves. Double that quantity will take another 2 – 3 minutes only.

When the spinach leaves are reduced to a soft mound at the bottom of the pan, take the lid off, raise the heat and drive off the rest of the moisture. Now the spinach is termed *en branche* by the French and is ready for serving. All it needs is a little chopping and squeezing.

The mistake people often make is to continue to cook the leaves, when they will obligingly go on to exude more and more moisture, shrinking all the time as they do so. You can stop this by various methods. First, proceed as above then take the spinach off the flame and serve. Secondly, add some butter, cream or oil to coat the leaves. Thirdly, add flour and make the spinach into a roux preparatory to adding milk or cream for a spinach sauce. Sometimes spinach also needs to be coated in this way as a preliminary stage in a particular recipe.

Large spinach leaves once blanched are useful for wrapping and surrounding mousses and moulds. They can be used like vine leaves in *dolmades*. In fact, as a change from salad, the next best way of consuming raw spinach leaves is wrapped around other vegetable mixtures. These wrapped mousses and moulds can also be cooked. The spinach which is at the top of the mould must be protected by paper and foil; even so it will dry out and darken. Once turned out, the mould will have the base as the top, so it will not show. Cut these moulds like a cake, if large. They can also be made in individual portion sizes and these make excellent first courses.

Spinach Soup is one of the best of soups and it is the easiest to make. For six people you will need 900 g / 2 lb spinach and 2.25 litres / 4 pt of liquid, which is made up of half vegetable stock and half milk. The latter can be soya milk for a soup which is healthy and slimming (I enjoy the nutty taste of the soya) or cows' milk (or even a mixture of that and cream for added richness).

Simply cook the spinach leaves in the stock for about 4 minutes. Leave to cool, season, then liquidize in a blender. Add the milk, soy milk or cream and milk and reheat gently. Garnish with yogurt and chopped chives or a few green and pink crushed peppercorns, or geranium petals. Spinach soup is perfectly delicious eaten cold as well.

Because of its intensity of flavour, spinach is often teamed with cream, cheese and eggs as these tone it down and make a flavour that is rich but acceptable. One of the most famous dishes with spinach is *oeufs à la florentine*, *Eggs Florentine*. This dish, in which the spinach leaves or purée are mixed with a cheese sauce and poured over poached eggs, used to be a popular beginning to a meal in Italy or in Italian restaurants over here. Most of us would now be content to enjoy it as a main course or as a supper dish.

The success of the dish is the fusion of the uncooked yolk with the cheese sauce and the fragments of spinach beneath. If the yolk is cooked through, the dish fails. Thus, the dish becomes a good test for the quality of a restaurant kitchen. If the egg is poached perfectly – the white set, the yolk still runny – the next stage, placing the spinach and cheese sauce over the eggs to brown in the oven or under the grill, will succeed in cooking the yolk. The trick is to allow the poached eggs to get cold, then you can safely cover them with the topping and grill and the eggs will then warm sufficiently to be eaten, but will not start cooking.

There is another method that I sometimes follow. This is to serve the poached eggs on top of the spinach mixture, garnished with fresh sage and grated Parmesan. It is not the classic dish, but it works well and it ensures that perfect combination on the palate of raw yolk and cheesy spinach.

For this dish and a spinach soufflé or *timbale* there are again two methods of treating the spinach, leaving it *en branche* (see above) and just chopping it or turning it into a purée. It depends on you and your personal tastes. For myself, I would always leave it *en branche* because I like texture and variety in food.

Lastly, there is a famous recipe for *Buttered Spinach* quoted by Elizabeth David in *French Country Cooking*. It was originally given in a weekly magazine published in Paris in August 1905. The story concerned the Abbé Chevrier and his best friend, the celebrated gastronome Brillat-Savarin: the latter was intrigued by the spinach cooked in butter at the Abbé's table. 'Nowhere does one eat spinach, simple spinach cooked in butter, to compare with his.' Brillat-Savarin discovered the recipe. It is cooked over five days.

The spinach is cooked in the normal way and afterwards, for each 450 g / 1 lb pound of spinach, 115 g / 4 oz of butter is added over a low flame. Stir the spinach into the butter so that the spinach absorbs the butter. Put aside and keep in a cool place. The next day, add 45 g / 1½ oz of butter to the spinach over a low flame until the spinach has absorbed it, working the spinach with a wooden spoon into the melted butter. Repeat this procedure for the next two days and finally, on the day you are going to serve it, add another 55 g / 2 oz of butter. Each 450 g / 1 lb of spinach will absorb 300 g / 10½ oz of butter, and the spinach will dissolve into a velvet purée. Mrs David adds that it is advisable to cook 2 or 3 lb of spinach to this amount of butter.

I cooked this recipe once with about 900 g / 2 lb of spinach to 285 g / 10 oz of butter. It was astonishing, one of the most blissful few mouthfuls I have ever enjoyed. One feels rather that it is heart-attack time, however, yet for one dinner party it is very well worth trying.

SPINACH ROLLS

These are perfect appetizers or party snacks.

12 spinach leaves, blanched
for the filling:
250 g / 9 oz Ricotta cheese
generous handful of chopped herbs
(parsley, chives, dill, basil, coriander)
bunch of spring onions, chopped
1 tsp paprika
2 garlic cloves, crushed

Mix the filling ingredients together well. Place a dessertspoonful of the filling on the corner of each blanched spinach leaf and roll it up, tucking the ends in carefully as you go.

SPINACH CROQUETTES

900 g / 2 lb spinach, cooked en branche *(see opposite)*
2 tbsp gram flour, plus more for dusting
1 tbsp garam masala
1 egg, beaten
sea salt and freshly ground black pepper

Mix the cooked spinach with the gram flour, garam masala, beaten egg and seasoning. Leave in the refrigerator to firm up a little.

Form the firm mixture into cakes or croquettes, roll these in flour and fry briefly on either side until uniformly golden.

PURÉE OF SPINACH AND POTATO

This is excellent with roast game or meat.

450 g / 1 lb spinach, cooked and puréed in a blender
900 g / 2 lb floury potatoes, cooked and mashed
30 g / 1 oz butter
5 tbsp milk or cream
sea salt and freshly ground black pepper

Combine the potato and spinach purées and mix thoroughly by hand. (Do not use the blender as it turns the potato into glue.)

Add the butter, milk or cream and seasoning until you have a smooth green purée.

SPINACH ROULADE

This makes an excellent first course, and you have a choice of fillings.

450 g / 1 lb spinach
4 eggs, beaten
55 g / 2 oz Gruyère cheese, grated
55 g / 2 oz Parmesan cheese, grated
sea salt and freshly ground black pepper
for the Taramasalata filling:
170 g / 6 oz smoked cod's roe
juice from 1 lemon
1 garlic clove, crushed
5 tbsp olive oil, maybe less, maybe more
for the Red Pepper Cream filling:
250 g / 9 oz Mascarpone cheese
1 red pepper, deseeded and diced small
1 red onion, diced small
1 tsp paprika

First make the roulade: cook the spinach, then chop it small. Add the beaten eggs, Gruyère and season.

Preheat the oven to 220°C/425°F/gas7. Line a roulade or Swiss roll tin with greaseproof paper and butter or oil it well.

Pour the spinach and egg mixture into the lined tin and spread it out evenly. Bake in the oven for 15 – 20 minutes, until it is crisp at the edges and spongy in the centre. Take out and allow to cool a little.

Have ready another sheet of greaseproof paper sprinkled with the grated Parmesan, upturn the roulade on to this and peel away the piece of greaseproof lining paper. Once cool, the roulade is ready to be filled and rolled.

To make the Taramasalata filling: cut the roe up with a knife, skin and all, place in a powerful blender and add lemon juice and garlic. Blend well, then start to add the olive oil in a steady stream – the result should be thick, creamy and smooth.

It is impossible to give exact measurements for the amount of oil as it depends on the quality of the roe. Some will absorb much more oil than others. Sometimes the roe will curdle, in which case a dash of whisky or brandy will help; otherwise you have to start again.

To make the Red Pepper Cream filling: simply mix the cheese with the diced vegetables and season with paprika.

Smooth either the taramasalata or the red pepper cream over the roulade, leaving a 1 cm / ½ in space around the edges. Roll the roulade gently from one end to the other, using the piece of greaseproof paper under it to help you get started. Place the roll on a platter, seam side down and chill for an hour or two.

Serve cut across into 1 cm / ½ in slices.

ABOVE Spinach, pine nut, papaya and avocado salad and spinach roulade

SPINACH, PINE NUT, PAPAYA AND AVOCADO SALAD

This simple and refreshing dish makes a terrific colourful first course or it can be served for luncheon with some good crusty bread.

675 g / 1½ lb fresh spinach leaves
4 tbsp olive oil
1 tbsp wine vinegar
pinch of sugar
1 ripe avocado
1 ripe papaya
3 garlic cloves, sliced
55 g / 2 oz pine nuts
1 tbsp chopped tarragon
sea salt and freshly ground black pepper

Wash, trim and drain the spinach leaves, then pat them dry in clean tea towel.

Make a dressing with 3 tablespoons of olive oil, the vinegar, sugar and seasoning.

Peel and slice the avocado into a large salad bowl and pour the dressing over it. Toss so that the avocado slices are well covered. Do the same with the papaya fruit, adding it to the dressing.

Pour the last tablespoon of oil into a pan. Add the sliced garlic and the pine nuts. Fry for a moment or two so that they turn golden brown. Leave to cool.

Add the spinach leaves to the salad bowl and toss thoroughly. Pour over the pine nuts and garlic with its oil and serve garnished with the chopped tarragon.

SWISS CHARD

Beta vulgaris subsp. *cicla*

Chard is very like spinach in flavour, but the stalks are eaten as well. Some people prefer it to spinach and pies made from it using the stems as well as the leaves have a pleasant silky richness. Also called leaf beet, spinach beet, seakale beet and white beet, chard is a form of beetroot grown for its leaves rather than its root. How it came to be called Swiss is somewhat of a mystery. The word chard comes from the Latin and French words for thistle. The French word for cardoon – which is a type of thistle – is '*chardon*' and a seed merchant may well have given chard the extra title of 'Swiss' to distinguish it from the cardoon.

The ancients had no such problems; they valued the beetroot for its leaves and spoke of white, red and black varieties. If the root of the beet was used it was for medicinal purposes. The wild form is found in the Canary Islands, in the whole of the Mediterranean region as far as the Caspian, in Persia and Babylon as well as around the coasts of Britain.

In 1597 the herbalist Gerard notes, 'the common white Beet hath great broad leaves, smooth and plain: from which rose thicke crested or chamfered stalks ...' He had also heard of another sort, 'red in colour, both in root and stalk, full of a perfect, purple juice'. He recommends using the leaves in winter as a salad, dressed with vinegar, oil and salt. He expresses doubt about the red and beautiful root, but has been assured that it is good and wholesome.

We, of course, now eat the root from one plant and the leaves from another, but we do tend to live wasteful lives. When I lived on the Greek island of Lesbos in the 1960s, in the early spring beetroots were one of the few fresh vegetables to buy. We boiled the roots and ate the leaves, first simmered briefly then dressed with Greek olive oil, vinegar and salt. Excellent they were too.

NUTRITIONAL AND MEDICINAL

Swiss chard contains less oxalic acid than spinach, so its nutrient content – roughly similar to spinach – is probably more easily absorbed by the body.

VARIETIES, CHOOSING AND STORING

Swiss chard now comes in both white and red varieties. The latter looks sensational growing in the herb or flower border, as the stems are a bright military scarlet. Alas, this colour goes on cooking and the leaves turn a darkish green (see cabbage on page 91 for the reason for such pigmentation loss). This may well be what the Greek philosopher and botanist Theophrastus (370 BC) meant when he named a black or dark green variety. Red chard is called either ruby chard or rhubarb chard. There is even a rainbow chard that produces red, yellow, purple and white stems. All chards tend to be hardy; the leaves can be cut all the year round and the plant continues to send up healthy new shoots.

Chard keeps better than spinach, but it is best to cook it within two days. Wrap it in paper and store in the salad drawer of the refrigerator. Of course, the stalks keep better than the leaves.

PREPARING AND COOKING

The south of France, around Nice, is where they most value chard. American food writer Richard Olney tells us in *Simple French Food* that the white stems are enjoyed, while the green leaves are given to the rabbits or ducks. Certainly, the leaves once cooked are indistinguishable in flavour from spinach. The stalks, however, are another experience altogether. Trim and cook them in a little stock, simmering for about 5 minutes. Then serve them with lemon alone, as one might eat asparagus. They can also be served as a vegetable with the rest of the meal.

There is a Provençal recipe which is the main speciality of Nice and is served traditionally as the grand dessert at Christmas Eve, *tourte de blettes*. In a manner reminiscent of medieval spinach tarts, the Swiss chard leaves are used mixed with raisins, apples, pine nuts, lemon and cheese. An Auvergne recipe, *le pounti*, is a chard and ham flan which may also be flavoured with prunes and raisins. Strangely, Elizabeth David dismisses chard as having little flavour and in her *French Provincial Cooking* gives only one recipe.

RIGHT *White and ruby chard*

SWISS CHARD PIE

My version of the Niçoise *tourte* omits the fruit and sugar, so is not in the least traditional, but the advantage is that it uses the whole plant. I cannot imagine why stalk and leaf are not more often eaten together.

900 g / 2 lb Swiss chard
55 g / 2 oz Gruyère cheese, grated
55 g / 2 oz Parmesan cheese, grated
55 g / 2 oz sage Derby cheese, grated
300 ml / ½ pt single cream
2 eggs, beaten, plus 1 extra yolk
sea salt and freshly ground black pepper
for the shortcrust pastry:
285 g / 10 oz best-quality plain flour
140 g / 5 oz best butter
pinch of sea salt
2 – 3 tbsp ice-cold water
little milk, for glazing

First make the pastry (make sure you have both the flour and butter very cold). Sift the flour and salt into a mixing bowl and grate the butter into that. Rub in the butter until you have a mixture with the texture of fine breadcrumbs. Add just enough ice-cold water to bring the paste together into a ball. Wrap in greaseproof paper and chill for 30 minutes.

Preheat the oven to 190°C/375°F/gas5. Allow the pastry to come to room temperature (about 10 minutes out of the refrigerator) before rolling.

Chop the chard, both stalk and leaves. Throw into a heavy-based pan, cover and cook over a low heat for 6 – 10 minutes, until both leaf and stalk are tender, then drain thoroughly, squeezing out all moisture.

While the chard is cooking, prepare the pastry case: keeping back about one-quarter of the pastry for the lid, roll the rest of the pastry and use to line a spring-form pie tin with a diameter of 27 cm / 10¾ in and a depth of 5 cm / 2 in. Prick the pastry base with a fork, line with greaseproof paper and weight with

beans. Bake the pastry blind for about 5 minutes.

Place the drained chard in a bowl and mix in the cheeses, cream, the 2 whole eggs and seasoning. Stir, mixing thoroughly. Pour into the pastry case. Roll out the remaining pastry and put it in place over the pie, moistening the edges with a little milk.

Glaze with the remaining egg yolk beaten with a little milk. Protect the top with greaseproof paper or foil and bake for 45 minutes or so, taking off the greaseproof paper or foil for the last 5 minutes to brown. Serve the pie cold or warm.

ON PASTRY

We all want to achieve crumbly buttery pastry that melts in the mouth. When you need the pastry to make good sides which stand up to hold a deep pie, however, that kind of pastry tends to collapse, so it must be baked blind with plenty of dried beans wrapped in foil to hold the sides up.

For tarts, this is not a problem. It is the ratio of fat to flour, with the addition of the minimum of ice-cold water, which gives the crumbly texture. It is the quality of the fat and flour which gives the flavour. The best pastry I ever made used unpasteurized butter, stone-ground flour and well water. How's that for purity? Food can only be as good as its ingredients. You can dispense with the butter and substitute crème fraîche or fromage frais, but if you use low-fat kinds of either the pastry will be brittle and tasteless. Rich pastry needs a high amount of fat.

As tap water is now of such doubtful quality it is advisable to use still bottled water for your pastry.

CHARD DHAL

This is another recipe that uses both leaf and stalk and which is unexpectedly creamy.

55 g / 2 oz brown lentils
1 tbsp turmeric
450 g / 1 lb chard leaves and stalks, chopped small
115 g / 4 oz orange lentils
sea salt and freshly ground black pepper

Lentils do not have to be soaked. Into 1.75 litres / 3 pints of boiling water, throw the brown lentils with the turmeric. Let simmer for 20 minutes.

Now add the Swiss chard and bring back to the boil. Simmer for another 10 minutes. Now add the orange lentils and simmer for yet another 10 minutes.

Season with the salt and pepper. The mixture should have soaked up all the water and can be served at once, though it will do no harm to rest in the warm. The dhal should have an amazingly buttery texture and creamy flavour.

WILTED CHARD SALAD

Young Swiss chard leaves can also be used as a salad vegetable. Like young spinach, however, they are also excellent as a wilted salad. The term 'wilted' is somewhat off-putting, but all it means is that the leaves are thrown into a pan with a flavoured oil, spices and other ingredients and stir-fried for a second or two so that the leaves just reduce, or grow a little limp, and get coated with flavourings. These salads are supposed to be served warm, and jolly good they are too.

2 tbsp mustard or sesame oil
55 g / 2 oz ginger root, grated
3 garlic cloves, sliced
1 red chilli pepper, deseeded and sliced (or broken up if dried)
1 tbsp each of crushed cumin seeds, cardamom and coriander
450 g / 1 lb Swiss chard, leaves torn and stalks chopped (keep separate)
¼ tsp sea salt
¼ tsp caster sugar
1 tbsp amchoor (dried mango powder), or more to taste

Heat the oil in the pan and throw in the ginger, garlic, chilli, spices and chopped chard stalks. Stir-fry until the spices darken and the garlic browns; by then the chard stalks should just begin to soften.

Now add the torn leaves and stir-fry rapidly for about 30 seconds. Add the salt and sugar, stirring again, then tip the pan contents into a salad bowl, sprinkle with the amchoor and serve immediately.

BEETROOT

Beta vulgaris subsp. *vulgaris*

We tend not to value this root, relegating it to cold salad or pickles. I wonder why, for its flavour is delicious and its colour remarkable. It makes the most stunning soup – *bortsch* – and when eaten small as a hot vegetable is an excellent foil to game. I suspect that the beetroot's habit of oozing its cardinal red dye over everything may be a deterrent to its popularity. There are, however, ways and means of guarding against this and they should be taken so that we can enjoy the vegetable more.

We know that the Greeks merely ate the leaves, but the root then was treasured for its medicinal qualities. The Romans cultivated the root and began eating it at table. Apicius gives a recipe for a beetroot salad to be dressed with mustard, oil and vinegar.

Beetroot plants like salt and so they are a useful crop for growing on reclaimed land near the sea. The beetroot is important economically, for its siblings, the sugar beet and the mangel-wurzel, both played dramatic parts in recent history. Ever since the sixteenth century it was known that some roots contained sugar, but Europe's appetite for the sweet stuff used to be minuscule compared to nowadays. In 1788 the French were only consuming one kilo of sugar per person per year, while in America today the consumption is one kilo per person per week.

In 1747 a German chemist, A.S. Margraaf, isolated sugar from beets. The sugar beet looks like a large white parsnip, each one can be 900 g / 2 lb in weight and 30 cm / 1 ft in length. In 1776 the first factory to process sugar from beet began working and twenty years later another factory in Austria was opened by a pupil of Margraaf's. The process was not particularly efficient, however, only 2 to 3 per cent of sugar was extracted from the roots. It took war and a naval blockade for the sugar beet industry to take off in a big way. The English blockaded Napoleonic France, cutting the country off from its sugar cane supplies. Napoleon ordered 70,000 acres to be planted with sugar beet. In 1812 Benjamin Delessert, a French financier, opened a refinery in Paris to process the beets. The development of the sugar beet used the white variety and by 1880 beet sugar was more widely consumed than cane sugar in Europe, except for Britain. Production of beet sugar in Britain only began seriously in 1924 when the Treasury granted a huge subsidy to its manufacture.

The mangel-wurzel – for long a subject of rustic schoolboy jokes – is a reddish orange in colour, not unlike a swede, and is grown as winter fodder for cattle. Its name is merely German for beetroot; it had been grown in Germany and Holland ever since the 1650s and from there it spread to northern France where, in times of famine, it was eaten by people as well as cattle. Confusion between the German '*mangold*' (beet) and '*mangel*' (dearth) led the French to call it '*racine de disette*' (root of scarcity/dearth).

In the latter part of the eighteenth century, during periods of hardship in rural England (the Enclosures Acts had taken the common land away from the farm workers, so they could not graze their few animals and could not gather wood for fires), the mangel-wurzel was planted to provide cheap food. Its name and reputation were against it, however, and people were convinced it did not save them from famine but brought it and the stigma of it to their land. It was, they thought, a food for cattle and would remain so. Already cattle were being fed through the winter with turnips and oil seed rape cakes (after the oil had been pressed out of the crop), but it was not until the next century that mangel-wurzels were grown for cattle.

They still are, around me in East Anglia. In the autumn months, great piles of mangel-wurzels lie in the fields waiting to be carried to the farms. Cattle love them as they too like sweetness.

NUTRITION

The green leaves are very high in vitamin A and therefore make a good substitute for spinach. In fact they have more iron, calcium, trace minerals and vitamin C than spinach itself. The roots are rich in potassium and fibre but also contain oxalic acid (see page 45).

VARIETIES

Beetroots come in all shapes: round, ovoid or long and tapering like a carrot. The taste is the same whatever the shape. There is also a white beetroot which tastes exactly the same as the red, but somehow a beetroot the colour of potato is off-putting.

Round red varieties – much the most popular – include Boltardy, Detroit and Monopoly, all of which are bolt-resistant. For a small beetroot, plant Detroit-Little Ball, which is used for commercial pickling. Of the yellow varieties, the best is considered to be Burpee's Golden which comes from the USA and has yellow flesh which does not bleed when cut. A good white variety is Snow-white or Albina Vereduna; good long varieties include Long Blood Red or Cheltenham Green Top and cylindrical varieties include Furono and Cylindra.

CHOOSING, PREPARING AND STORING

Buy your beetroots small if possible. Also buy them as fresh as you can and with their leaves attached, so that you can see how fresh they are. The leaves will sag after the first day.

Beetroots store well, up to a few weeks if necessary. After that they will certainly start to get soft and will not be worth cooking.

To prepare beetroots: wash the earth off them without breaking the skin, then twist or cut the leaves off leaving about 10 cm / 4 in of stalk (otherwise the colour will run) and don't cut off the tapering root.

COOKING

Boil beetroots for about 30–40 minutes. Larger and older roots can take 60 minutes or more to cook through. Drain, leave to cool a little, then rub off the skin. If you want your baby beetroots very hot you will have to impale them on a fork (when they will bleed a little) and peel them that way. Older, larger beetroots can be baked in the oven either just placed in an oven dish with the lid on or wrapped in foil. They will cook in a medium oven (190°C/375°F/gas5) and take up to one or two hours, depending on the size of the root.

One of my most favourite meals is a collection of young new-season vegetables – beetroots, tiny potatoes and carrots, turnips, beans and onions – all just simply boiled or steamed and then served on a great platter with a large accompanying bowl of glistening *aïoli* mayonnaise. It is one of the really great treats of early summer.

My mother used to serve a white sauce with hot beetroots. This was satisfactory except that the sauce slowly turned pink in splodges, which can be unsightly. Butter or a garlicky vinaigrette can also be served with hot beetroot, as well as a sweet-and-sour sauce made with brown sugar, chilli, vinegar and oil.

This is a recipe from the classic French work *Le Cuisinier Royal* (1682) which shows how imaginative they were with beetroot then: 'First rost your Beet Roots in the Embers and peel them very well. Cut them into pieces and give them a boil with a piece of sugar, a little salt and cinnamon and make them like marmelade, and put them into a fine paste with some green Citron rasp'd and a piece of butter and do not cover it but when it is baked, serve it away with perfumed sugar and Orange Flowers.'

Raw grated beetroot makes a marvellous salad and I can't imagine why it is not eaten more often. Consuming beetroot will, of course, turn your urine pink or red: it is quite harmless, so do not worry.

Grated raw beetroot can be dressed with oil and vinegar and have other raw vegetables mixed in with it: diced tomato and onion, or finely sliced leek and grated cabbage. It is particularly good with chopped dried fruits, apricots, prunes and raisins. I cannot recommend these salads more highly.

One of the great Eastern European taste combinations is horseradish and beetroot. We cannot usually buy fresh horseradish in order to make our own, though we can grow it, so one has to rely on commercial brands; these are rarely any good but if you find one you like, add a spoonful of horseradish to any of the beetroot salads (the beetroot dressed with sweet-and-sour sauce certainly could also benefit).

The flavour of beetroot – that sweet earthiness – needs very little to accentuate it. This flavour is experienced most powerfully in the soups made from the roots, with their colours of deep garnet and amethyst they belong to soup royalty. Jane Grigson enthuses and makes the point: 'I do not understand why the English and Italians, and even more the French who understand such things so well, have not developed some striking beetroot soup of their own'.

ABOVE Grated beetroot salad with dried fruits, roast beetroot wedges and beetroot moulds

BORTSCH

for 6

450 g / 1 lb small beetroots
2.25 litres / 4 pt vegetable stock (see note below)
sea salt and freshly ground black pepper
sour cream or yogurt, to serve

Peel the beetroots and slice or dice them into chunks. Add these to the stock and bring to the boil. Simmer for about 45 minutes. Allow to cool. Blend to a purée. Season and chill until needed. Serve chilled or hot with the sour cream or yogurt in a bowl.

VARIATIONS

- Add a couple of onions and some crushed garlic.
- Instead of blending, crush the cooked beetroot with a potato masher. Leave to get cold, then strain. You will have a lighter soup, more akin to a beetroot drink.

BEETROOT MOULDS

This is a favourite summer first course, as their taste and visual charm are so enticing. To the strained beetroot stock of Bortsch, variation 2, add 2 tablespoons of agar agar, gelozone or gelatine and heat until thoroughly melted. Pour into moulds. Chill until set.

Dilute some light fromage frais, smetana or crème fraîche with a little white wine or dry sherry and flood some chilled plates with this. Turn the moulds out on the plates so that they sit on the centre.

Garnish with a little chopped chives, viola petals, salad burnet, chopped lovage, geranium petals – anything with flavour that also looks striking.

For a darker version flavoured with ginger, cook the beetroots again as for Bortsch but add 55 g / 2 oz of grated root ginger. Blend together in the stock, let cool a little and then add the gelling agent.

ORACHE

Atriplex hortensis

This beautiful plant, also called mountain spinach, is either red- or golden-leaved. It seeds itself each year in my own garden and flourishes. It is tall – up to 2 m / 7 ft – and has small triangular pointed leaves: when young these can be eaten in a mixed salad; the larger leaves can be treated like spinach. The plant is so beautiful, however, I can never endure to harvest it.

The name orache is a corruption of *aurum* (gold), because the seeds mixed with wine were supposed to cure yellow jaundice. Also, the seeds heated with vinegar, honey and salt and then applied in a poultice were used as a cure for gout.

Its history as a kitchen vegetable is a long one. Its cousin sea orache (*Atriplex halimus*) is one of the very few plants indigenous to Egypt that can sustain man. Dioscorides, the first-century Greek physician, mentions it being cooked and eaten. It grows around the shores of the Mediterranean and northern Europe.

Orache was known by the Greeks, who boiled the leaves; the Romans knew it as *atriplex*. The wild form of orache (*Atriplex patula*) was loathed by the sixteenth-century herbalist Gerard. who calls it stinking: 'It groweth in the most filthy places that may be found.' He goes on to say that its smell is like stinking fish or the rammish male goat, the common term for it being 'stinking mother-wort'. The seventeenth-century herbalist Culpeper is in agreement, but describes its use in curing all manner of women's diseases. Perhaps this twinning of the female with a repugnant herb reflects more on contemporary male attitudes to women than to anything else.

The common orache is called 'fat hen' in some parts of Britain and, indeed, it belongs to this species that grows in farmyards and fields on muck heaps. The *Reader's Digest Field Guide to the Wild Flowers of Britain* entirely ignores what Gerard and Culpeper said of the common orache and comments: 'The garden form of orache (or orach) was served until quite recently as a table vegetable resembling spinach.' Its entry under '*Atriplex patula*', however, seems to confuse the common and garden varieties , and claims that we treat them both as a weed.

Garden orache was eaten in the Ancient World much as other leaves like mustard, dock, nettle and mallow were – the young leaves eaten raw; the larger, older leaves boiled. Apicius gives a recipe where he purées the nettle leaves and then adds eggs. At the same time in Britain they were eating much the same mixture; mustard and orache seeds were found in the intestinal areas of the skeletons found at ancient Glastonbury. The family has always been an important food plant for man. The seeds contain fat and albumen and they formed part of the last meal of the preserved Iron Age Tollund Man.

As the plant is easy to grow in temperate climes, one can assume it remained throughout the dark and medieval ages as part of the food of the majority of people. Later it was much used in seventeenth-century soups and stuffings, being obliging in that once cooked it reduces itself to a purée. John Evelyn comments: 'Being set over a fire, neither this nor the lettuce needs any other water than their own moisture to boil them in.' One can imagine it as a part of the range of herbs and vegetables used at medieval banquets, as well as being an ingredient for the soup of the poor. We next hear of it in 1538 when William Turner, the physician, botanist and Dean of Wells Cathedral, refers to orache in his *Herbal*. From then on it is continually mentioned by herbalists. By 1806 three different kinds of orache are listed in the *American Gardener's Calendar* by Bernard McMahon.

Sadly orache disappeared from our gardens this century. However, the seeds of both the red orache (*A. hortensis rubra*) and the gold orache (A. *hortensis*) can be obtained from specialist suppliers (see page 279).

GOOD KING HENRY

Chenopodium bonus-henricus

This is another plant that has disappeared from our gardens, but which was valued in the past. Writers disagree as to its culinary charms. Geoffrey Grigson in *The Englishman's Flora* (1975) writes: 'Since the young shoots and flowering tops boiled and eaten

ORACHE *Atriplex hortensis*

with butter are neither very pleasant nor unpleasant, this plant hardly lives up to its name as an old pot-herb.' English food and garden writer of the 1940s and 50s, T. A. Layton has a quite different view: 'Here is a vegetable which is quite extraordinarily easy to grow and tastes every bit as good as spinach. When the shoots are young and are earthed up in the early spring, it is as good as asparagus...' I have grown it for many years and the trick is to do as Layton says. Do not bother with the flowering tops or the leaves, though they taste much like spinach.

The name Good King Henry is only vaguely to do with King Hal; in fact, the original German had nothing to do with a monarch at all. The plant got its name from Tudor herbalists who translated the German *Güter Heinrich* (good Henry) to distinguish it from *Boser Heinrich* (bad Henry), a poisonous plant (*Mercurialis perennis*) which it resembles. The Germans did not call it 'King' and the Heinrich was probably an elf in the same way that we give many of our plants the title Robin, after the malicious fairy of medieval folklore, Puck or Robin Goodfellow. Tudor herbalists, delighted with the title 'good Henry', decided that a little royal flattery might not be a bad thing.

The plant was certainly cultivated with enthusiasm in the sixteenth century and it remained popular in parts of England, Lincolnshire and the Midlands, though it seemed to have died out elsewhere. In 1783 Henry Bryant, the botanist, writes: 'Formerly cultivated in English gardens but of late neglected, although certainly of sufficient merit.'

H. G. Glasspoole in his history of our common cultivated vegetables for the Ohio State Board of Agriculture in 1875, commented that in Lincolnshire 'it was preferred to garden spinach and the young shoots used to be peeled and eaten as asparagus.'

MARSH SAMPHIRE or GLASSWORT

Salicornia europaea

This is the samphire that you will now find on sale in enterprising fishmongers, coming into season in August and lasting for a month or perhaps two.

It grows in estuaries and salt marshes all around the European coastline, appearing in June. A walk at low tide will show the bright green spears rising about 7.5 cm / 3 in above the mud. You can start picking them then, snapping them off above the root, or wait until August when most of the plant shows. Then the whole plant, root and all, can be gathered.

In the sixteenth century Gerard describes 'Glasse Saltwort' thus: 'Glassewort hath many grosse thicke and round stalks a foot high, full of fat and thicke sprigs, set with many knots or joints, without any leaves at all, of a reddish greene colour: the whole plant resembles a branch of Corall: the root is very small and single.'

Marsh samphire belongs to the branch of the Chenopodiaceae that contains plants, such as *Salsola kali* and Spanish *Salsola sativa* among others, which are particularly rich in soda, hence they were used in the process of making both soap and glass. There was a thriving trade from the Mediterranean, sending the ashes of these plants called 'barilla' to northern Europe for the production of glass.

Cattle love the salty flavour of marsh samphire and will devour it greedily. Sir Thomas More (1478) names it in a list of plants that would improve 'many a poor knave's pottage...glasswort might afford him a pickle for his mouthful of salt meat.'

Samphire needs very little cooking. It is generally sold in bundles and well washed: if not, rinse it under a cold tap, then blanch it by pouring boiling water over it. Leave it to soak for a few minutes, then drain and serve. Most writers on samphire recommend boiling it for much longer: my much respected friend and expert on wild food Richard Mabey actually says 10 minutes. I have tried steaming for 5, and found it slightly overcooked, and boiling for 3 – still overcooked. So I will stick to my blanching method.

How to eat samphire: pick the pieces up by the root, place the stem in the mouth and, with the teeth, pull off the green flesh from the rather wiry interior stem. Young samphire picked before August has no wiry centre. It is particularly good served with fish, and a whole bass or hake surrounded by a bed of samphire is a sight to inspire rapture.

QUINOA

Chenopodium quinoa

This is a very high-protein grain, packets of which you can now buy in health food shops. You can also grow it and harvest it in your own garden if you wish. It is a half-hardy annual and it will need a rich well-manured soil. If planted in the first week of May, it will grow 2-2.5 m / 7 – 8 ft by August. Its heads of seeds are a variety of colours: purple, pink, green, cream, gold and ochre (although the plant in its wild state has only dark seeds) and quinoa is a spectacular addition to the garden even if you don't harvest it.

Like all this family, the leaves will cook and taste like spinach. It is indigenous to the Pacific slopes of the Andes and was a staple food for the people of Peru and Chile. Garcilaso de la Vega, one of the great Spanish chroniclers of the sixteenth century (he was the illegitimate son of a Spanish knight and an Inca princess) records in the 1560s that the grain was called quinoa by the natives of Peru but *mujo* by the Spaniards. He writes: 'Both the Indians and the Spanish eat the tender leaf in their dishes because they are savoury and very wholesome. They also eat the grain in the soups, prepared in various ways.'

Other explorers noticed how savoury the seeds were when boiled in milk and then made into a type of gruel seasoned with chilli pepper and salt. At other times the grains were toasted, then boiled in water and strained; the resulting brown-coloured broth was seasoned and drunk as a favourite refreshment of the ladies of Lima. The seeds were also roasted and ground, then made into bread or mixed with fat and flavourings, rolled into balls and steamed.

Aizoaceae

NEW ZEALAND SPINACH

Tetragonia expansa

This plant was first recorded by Sir Joseph Banks in 1770 at Queen Charlotte's Sound, New Zealand. Two years later it reached Kew gardens. It thrives in dry hot regions and tends to be destroyed by a hard frost, but I grow it each year very successfully.

It was not until Captain Cook's second voyage around the world that, back in Queen Charlotte's Sound, the botanist Foster thought the leaves of the plant much resembled orache and wondered whether it too might be an antidote to scurvy. The sailors ate it and found that it did give protection against the illness caused by a deficiency of vitamin C.

The plant was slow to catch on in Europe: in England it began to be cultivated in 1821 and a little later in France. It appeared in seed catalogues in the USA in 1828. Few people grow it now, however, though – with its creeping tendrils, spear-shaped leaves and large seed pods – it is an attractive plant. The leaves must be picked young. These are the only edible part; the stalks are fibrous and the seed pods quite hard.

The flavour is very similar to spinach, but if the leaves are picked young they have a creamy consistency which is particularly delicious and, like sorrel, they almost purée themselves.

BELOW *New Zealand spinach*

Compositae

THE LETTUCE FAMILY

LETTUCE *Lactuca sativa*
CHICORY *Cichorium intybus*
ENDIVE *Cichorium endivia*
DANDELION *Taraxacum officinale*
GLOBE ARTICHOKE *Cynara scolymus*
CARDOON *Cynara cardunculus*
JERUSALEM ARTICHOKE *Helianthus tuberosus*
SALSIFY and SCORZONERA
Tragopogon porrifolius and *Scorzonera hispanica*

Polygonaceae

RHUBARB *Rheum rhaponticum*
SORREL *Rumex acetosa*

Portulacaceae

PURSLANE *Portulaca oleracea*

The lettuce family is one of the largest in the plant kingdom, with many flowers which are familiar to us – asters, dahlias, marigolds, zinnias – and the weeds, dandelions and thistles. It numbers many food plants, including sunflowers and artichokes, as well as all those green leaves we love to eat as salads and one of the most pungent herbs, artemisia or wormwood, which is used as a flavouring in absinthe and vermouth. A characteristic of the family is that its members tend to produce a bitterish milk sap, which, in many cases, seems to have soporific powers.

Included in this chapter is the related knotweed family, Polygonaceae, which contains over 800 species, many with astringent flavours and even blistering properties. Yet among them are valuable food plants, including herbs, shrubs and trees, several of which used to be cultivated and have now gone wild. Also included here is the purslane family, Portulacaceae, which contains 500 species of herbs and small shrubs, native primarily to the Pacific coast of North America and southern South America.

LETTUCE

Lactuca sativa

The lettuce is the most astonishing plant. Not only must it be the most popular vegetable today – eaten all over the world, even by people who have little liking for it but who believe it to be the token 'green leaves' to balance a protein meal – but in the Ancient World it was considered both an aphrodisiac and its opposite – an anaphrodisiac.

Contradictions dog the history of the lettuce. It was eaten either before or at the start of the meal to awake the appetite, or at the end of the meal because it was believed to be soporific.

Herodotus notes that lettuce appeared on the royal tables of the Persian kings around 550 BC. It is also listed among the 250 plants growing in the hanging gardens of Merodach-Baladan, King of Babylonia from 721 to 710 BC, who died in 694 BC. The earliest mention of the lettuce, however, is the giant cos lettuce – some nearly a metre tall – of the Egyptians, which appeared on many of their reliefs and which was sacred to the God Min.

The Egyptians believed lettuce was an aphrodisiac, suitable for the god of procreation and vegetation, because the milky sap from thick-stemmed lettuce seemed similar to semen. On the reliefs, the lettuce offerings which Rameses II gives to Min are phallic-shaped and Min himself sports an erect penis and appears to caress the huge lettuces behind him. Lettuce oil was used by the Egyptians to soothe the skin and to relieve headaches, and much of the lettuce was grown to harvest the seed for this oil. The oil was also used to anoint the salad itself.

In the rest of the Mediterranean, however, the lettuce was thought to be an anaphrodisiac, especially by the Pythagoreans, who meditated upon the spirit and valued the plant as a sedative on the flesh. After all, when Adonis died, Venus threw herself upon a lettuce bed to lull her grief and cool her desires.

As laudanum is an extract from lettuce sap, it is certainly true that lettuce has a certain soporific effect. One that Beatrix Potter was familiar with when she told the tale of Peter Rabbit. Greeks and Romans liked to add other leaves like rocket, which would have a stimulating effect, to their salad and counter the effect of lettuce.

The only lettuce with which the Ancient World was familiar was the cos, but there seem to have been many varieties; from the giant Egyptian types of cos to others with so broad a stalk that wicker gates of kitchen gardens were made from them; there was even a squat-growing plant called a Spartan lettuce. There were also black lettuces, as well as red, purple, crinkly, Cappadocian and Greek. One particular lettuce was named the 'eunuch' (because of its 'potent check to amorous propensities', Pliny tells us), while another called the 'poppy lettuce' had abundant juice and was particularly soporific.

It is Pliny who tells us of the Emperor Augustus being cured of an illness by his doctor, Musa, who gave him lettuce. So grateful was the Emperor that he had lettuce pickled in honey vinegar, thus ensuring that he could eat it throughout the year. Augustus also ate his lettuce at the beginning of the meal, but by the time of Domitian the trend had reversed and it was eaten at the end. Gerard quotes Martial as asking 'Telle me why lettuce, which our Grandsires last did eate/Is now of late become, to be the first of meat?' Gerard goes on to say that now it is eaten at both times, 'before the meal it stirreth up the appetite and afterwards it keepeth away drunkenness'.

In France and the USA, the cos lettuce today is

referred to as 'romaine'. This is because it was thought to have been brought to France by the Papal court's move to Avignon. The round-hearted lettuce was bred in monastery gardens some time in the early Middle Ages. Certainly the Goodman of Paris had much to say to his young wife on cultivating the lettuce: 'note that they do not linger in the ground, but come up very thickly, wherefore you must root them up here and there, to give space to the rest that they crowd not'.

In 1597, Gerard describes only eight different types of lettuce and by 1822 there were only thirty. It was almost invariably eaten raw, though from Apicius onwards there are also some recipes.

VARIETIES

There are now hundreds of different varieties of lettuce. We have also added to the cos and round or cabbage types the crisp-head and loose-leaf types. The latter has the advantage for the gardener of allowing the leaves to be picked without uprooting the plant.

Of the cos type, the most popular commercially (for it is sold throughout the year) is Little Gem; next in taste and form must come Winter Density. Larger lettuces are Lobjoits Green, Paris White, Barcarolle and Rouge d'Hiver. The best of the round type of lettuce are Buttercrunch, Tom Thumb and the crisp-head types are Webb's Wonderful, Avoncrisp, Marmer, Regina dei Ghiacci and Rougette du Midi. Loose-leaved lettuces are Red and Green Lollo, Red and Green Salad Bowl and Ricciolina di Quercia.

NUTRITION

Lettuce is high in iron, calcium, phosphorus and potassium, and it has vitamins A, C and E. It is, therefore, high in antioxidants, so it is true that a fresh green salad every day is a necessary healthy part of the diet. Also, as vitamin B12 is created by bacteria acting on algae, it is a good thing to eat plenty of green leaves from the garden which have not been too scrupulously washed. It is, in my opinion, a myth that vegetarians and vegans will suffer from any B12 deficiencies when they consume so much greenery. However, sterilized polythene-wrapped leaves from the supermarket will not give B12.

BUYING, STORING AND PREPARING

If you can, buy all your lettuce and salad greens from local farmers, the farm shop or the greengrocers, where you know from whence the produce has come. In this way you can check whether the farmer uses pesticides or nitrate fertilizers. There is no doubt that organically grown vegetables have more flavour than any others and your own vegetables grown in your own garden are by far the best. The fresher the vegetable and the shorter the time it has left the earth, the greater the nutritional content as well as flavour.

The freshness of a lettuce is easy enough to see. It should be vibrant in colour and erect in leaf. Lettuces with yellowy leaves should not be on sale. If picked fresh, all salad leaves should be eaten that day or the following. If they must be stored, take them out of their polythene bag and keep in the salad drawer of the refrigerator or somewhere equally cool and dark.

On the whole, I am against washing vegetables and certainly the washing of salad leaves seems to me to be overdone; it is so difficult to dry them adequately even though salad spinners have now become a vital part of kitchen life. I would prefer to inspect the leaves and if dirt is found, remove it by discarding the leaf or wiping it clean. I like to feel that the salad leaves are as dry as possible to receive the dressing.

Never cut a lettuce leaf – simply pull it apart – and never shred it as some American books tell you. The Greek salad (and in Malta too) is built up on sliced lettuce, but it is eaten seconds after being prepared and this alleviates the harm. Exposing cut surfaces of the leaves breaks down the cellular structure, so that it grows limp quickly, loses nutrients and oxidizes quicker – the lettuce will then darken and bruise. Salad leaves should be treated gently, with infinite respect, as if they are a luxury item of food.

MAKING SALADS

You will need a large salad bowl, so that the leaves can be gently tossed without them being thrown out. Estimate with care the amount of dressing needed for the amount of salad. Each leaf should be finely coated, but there should be no excess dressing left at the bottom of the bowl.

Make the dressing in a small bowl or jug, or even in the salad bowl itself where it can be mixed or beaten strenuously, then *poured out* into a jug so that it can be added to the leaves after they are in the bowl. Use one part of vinegar or lemon juice to five of oil and add sea salt, a pinch of caster sugar and a good turn of the pepper mill, before adding everything else. There is a variety of possible additions, beginning with a crushed garlic clove, soy sauce, Worcestershire sauce, Tabasco, lemon or orange zest, finely chopped onion or shallot, raw or pounded boiled egg yolk, and many others – be adventurous and experiment.

The classic addition to the green salad in the nineteenth century in France was the *ravigote*, literally meaning the 'pick-me-up' (from *ravigoter*, to cheer or strengthen). This is an assemblage of four herbs – tarragon, chervil, chives and burnet. All of these were chopped very finely and placed in four separate saucers for the chef to create his particular mixture for that night.

COOKED LETTUCE

Lettuce needs very little cooking, but it is well worth doing occasionally, for the flavour is different and very pronounced. In my opinion, one of the greatest lettuce dishes is the French recipe given below for cooking older peas, Lettuce Clamart, given by Jane Grigson in her *Vegetable Book*.

Another quick recipe uses halved Little Gem lettuces, smeared with a sauce and grilled for about 2 minutes. For the sauce, try mixing about 30 g / 1 oz of blue cheese with a couple of spoonfuls of fromage frais. This is smeared over the inner hearts, which are then placed in a grill pan and cooked for about 2 minutes. Alternatively, mix a finely diced small green pepper with a teaspoon of fromage frais and use that instead of the blue cheese sauce to coat the lettuce.

Cos lettuces are often used for cooking, for they have enough vigour to be cooked and not wilt entirely. The round-head lettuces would decline rapidly into soft rags, but both Iceberg and Webb's Wonderful are excellent lettuces for an instant's stir-frying to make wilted salad. Those older recipes for braising lettuces in butter for 45 minutes, as an accompaniment to roasted joints, seem to me not worth doing. The flavour is bitter, the texture limp and slimy.

There is one other use for large cos lettuce leaves, to replace vine leaves in a type of LETTUCE DOLMADES. Here the central spines of the leaves are cut out and the leaves are blanched for 10 seconds, then drained thoroughly. A mixture of soy sauce, curd cheese, chopped onion and herbs (though it could be highly seasoned fish or minced game mixed with rice or bulghar wheat, moistened perhaps with a tiny amount of fromage frais) is thoroughly mixed and cooled. Then a heaped teaspoon of the mixture is placed at the end of each leaf and the leaf is rolled, tucking in the sides as you go. These refreshing *dolmades* make excellent appetizers or starters.

LETTUCE CLAMART

Clamart is the district in Paris where the best peas were grown. Every summer this becomes one of our most favourite luncheon dishes. No added water is necessary and, though it might look like a soup, you need to eat it with a knife, fork and spoon. It is a perfect recipe for summer vegetables. If you are lucky enough to have some baby artichoke hearts, they can be added too.

30 g / 1 oz butter
1 tbsp olive oil
1 onion or 3 shallots, finely chopped
2 or 3 large cos lettuces
675 g / 1½ lb fresh garden peas, shelled
4 or 5 baby carrots, diced
4 or 5 baby courgettes, thinly sliced
large sprig of mint
good pinch of sea salt
pinch of caster sugar

Melt the butter with the oil in a large saucepan. Throw in the onion or shallots, followed by the leaves from the lettuces. Lay them crosswise in the bottom of the saucepan to provide a nest for the vegetables.

Add the peas, carrots and courgettes. Then bury the mint among them, sprinkle salt and sugar over the lot and fit a lid tightly over the pan. Simmer over a low heat for about 15 minutes, when all the vegetables should have steamed in their own juices.

You will need soup bowls to eat the dish and a large pepper mill on the table, and possibly some more butter and olive oil.

ICED LETTUCE SOUP

30 g / 1 oz butter
1 onion or 3 shallots, finely chopped
2 or 3 large cos lettuces
1.1 litres / 2 pt vegetable stock
300 ml / ½ pt dry white wine
300 ml / ½ pt buttermilk or smetana
sea salt and freshly ground black pepper
chopped chives, to garnish

Melt the butter in a large saucepan and throw in the onion or shallots, followed by the leaves torn from the lettuces. Season lightly and place a tightly fitting lid on top.

Let cook over a low heat for about 10 minutes, then add the vegetable stock and cook for a further 10 minutes.

Let the pan cool, then blend the contents, adding the wine and buttermilk or smetana.

Taste and adjust the seasoning, if necessary. Chill in the refrigerator for 2–3 hours before serving sprinkled with some chopped chives.

LETTUCE SERVERS

Both Webb's Wonderful and Iceberg lettuces have recently come into their own as receptacles for other foods, especially in Chinese cooking. Minced quail, pigeon or duck is mixed with herbs and spices and served in a bowl with a pile of crisp lettuce leaves and a separate bowl of dipping sauce. The diner places some minced mixture upon a leaf, sprinkles a little sauce over it and wraps it around the food ready to eat. I find this particularly agreeable, and the crunch of the raw lettuce around a highly seasoned filling makes this a dish worth trying at home. The following recipe is my vegetable alternative.

2 Iceberg or Webb's Wonderful lettuces
for the filling:
170 g / 6 oz rice cooked with a pinch of saffron
85 g / 3 oz tabbouleh
2 onions or shallots, finely chopped
1 red pepper, diced small
2 fresh green chilli peppers, diced small
generous handful each of chopped parsley, chives and basil
1 tbsp fromage frais
1 tbsp soy sauce
sea salt and freshly ground black pepper
for the dipping sauce:
1 red chilli pepper, diced small
30 g / 1 oz root ginger, peeled and grated
1 tsp caster sugar
1 tbsp Dijon mustard
4 tbsp light soy sauce
1 tbsp dry sherry
1 tbsp sesame oil

First make the sauce: place the chilli and ginger in a bowl, add the sugar and mustard, then add the remaining ingredients. Stir well and let stand for several hours.

Make the filling by mixing the ingredients together thoroughly, then chill and place in a large bowl.

Separate the lettuces into their individual leaves (in the case of the Webb's the leaves come away entire with more ease).

A spoonful of the mixture is placed on the leaf and the dipping sauce is sprinkled over, then the leaf is rolled up and eaten.

VARIATIONS

You can add any of the following to the filling mixture: shrimps, prawns or scallops that have been grilled and diced small; flaked cooked fish (but not an oily fish); chopped cooked mushrooms or other fungi; chopped tomato flesh; artichoke bottoms or tiny sweetcorn; game which has been minced and moistened by stock (in this case the rice and tabbouleh amounts should be halved).

CHICORY

Cichorium intybus

The shoot or closed white bud, or *chicon*, is called by the French *endive*, and their *chicorée* is what we recognize as the curly-leaved endive or Batavia. This has always been confusing to travellers – French, British and American – but both these vegetables, however different in appearance they are, were developed from *Cichorium endivia* (see page 69).

The white bud kind, which was rediscovered by accident in the middle of the nineteenth century by a M. Brezier, is also referred to as Brussels chicory, Belgian endive or witloof, which is Flemish for 'white leaf'. M. Brezier was growing chicory roots for coffee and found some that had sprouted white leaves. He liked the taste and so did his family, and all his life he kept the cultivation method secret. When he died, his widow told the secret to M. Brezier's successor at the Brussels Botanical Gardens so that she could keep a little interest in the vegetable, which was becoming popular. After 1872 they were cultivated commercially. It was, and is, an easy vegetable to grow as it occupies little space and only needs warmth, water and complete darkness.

M. Brezier rediscovered these growing buds, for they had been cultivated in gardens in Germany in the sixteenth century. Before then, the Ancients certainly gathered chicory from wild plants. In thick undergrowth blanching can be a natural process.

It was not until the second half of the eighteenth century that the chicory root began to be used as an additive to coffee. A variety called Magdeburg chicory, which had a large tap root, was grown for this purpose. The root was roasted and ground before being added. As food manufacturers adulterated coffee with an overabundance of chicory, cheating was common. The addition of chicory to coffee was banned in Britain in 1832, but surprisingly the practice had collected its adherents who liked the taste and they protested. Thus, it was allowed as long as the labelling was clear as to the contents.

To get the characteristic tight sheath of white leaves, the chicory is forced by burying the roots in warm moist peat. If it is sold cosily nested in blue paper you can be certain it has been protected from light and will be less bitter. As this form of endive is such a new vegetable its cuisine is sparse, yet its potential is vast and it can be eaten raw or cooked.

VARIETIES

Both white and red varieties are available in the winter months. The latter is produced by Italy: one, Rossa de Verona, can be grown in the same way as witloof, the leaves being attractively tinged with red.

NUTRITION

Chicory is one of those plants like Jerusalem artichokes in which inulin comprises the major part of their carbohydrate. It is inulin which produces flatulence (see page 76). Chicory also contains vitamins A, C and K, calcium, phosphorus and potassium.

STORING AND PREPARING

Chicory keeps well, for up to a week in the salad drawer of the refrigerator or in a suitable cool and dark place. It is also easy to prepare: simply slice the root end away and discard any brown or damaged leaves. Some cooks claim that cutting chicory with a knife increases the bitterness (Elizabeth David recommends a silver or stainless-steel knife for cutting them), but I have never noticed any difference.

To eat them raw, the leaves need merely to be tossed in a vinaigrette, either on their own or with other leaves. A classic mixture is chicory and orange. Indeed, because of the leaf's slight bitterness, the sweetness of various fruits is a natural complement. Chicory with pink grapefruit, olives and roasted pine nuts is a good combination.

COOKING

Chicory is often baked, but it can be stir-fried or poached. Some people dislike the slipperiness of the cooked vegetable and this, it must be admitted, is a matter of taste. To prevent the droop of the cooked vegetable, do not slice the chicons, but cook them whole. If serving the chicory as an accompanying vegetable, allow one for each person; for a supper or luncheon dish, then allow two or three.

ABOVE *Stir-fried chicory*

POACHED CHICORY

6 or 8 heads of chicory, trimmed
300 ml / ½ pt vegetable stock
150 ml / ¼ pt sweet white wine
2 tbsp grated Parmesan cheese
sea salt and freshly ground black pepper

Pack the chicory, head-to-tail sardine-fashion, into a large saucepan and pour the stock and wine over, there should be just enough liquid to cover. Bring to the boil and simmer for 30 minutes.

Take out the chicory, draining well. Place in an oven dish and keep warm.

Preheat a hot grill. Reduce the cooking liquid to about 150 ml / ¼ pint and season carefully. Pour this over the chicory and sprinkle with the Parmesan. Grill for a few seconds to brown the surface lightly.

STIR-FRIED CHICORY

1 tbsp sesame oil
2 – 3 garlic cloves, sliced
30 g / 1 oz root ginger, peeled and grated
1 dried red chilli pepper, broken up
3 heads of chicory, separated into their leaves
1 tbsp peanut butter
pinch of caster sugar
pinch of sea salt

Heat the oil in a wok and throw in the garlic, ginger and chilli. Stir-fry for a minute.

Add the chicory leaves and stir-fry for another minute. The leaves should still be crisp.

Mix the peanut butter with 2 – 3 tablespoons of water to make a thick paste. Add this to the wok with the salt and sugar. Toss in the pan and serve.

SIMPLE BAKED CHICORY

Salting the vegetable before cooking makes it exude more liquid, so we season only after baking.

4 or 5 heads of chicory, trimmed
30 g / 1 oz butter
sea salt and freshly ground black pepper

Preheat the oven to 190°C/375°F/gas5.

Butter an ovenproof dish and pack the chicory into it. Dot the rest of the butter over, then cover the dish with foil and bake in the oven for 45 minutes.

Season the chicory and serve in its juices.

CHICORY IN A FRUIT GLAZE

2 tbsp strong fruit jelly (medlar, quince, sloe, damson, blackberry, blackcurrant or apricot)
30 g / 1 oz butter
4 or 5 heads of chicory
sea salt and freshly ground black pepper

Melt the jelly in 5 tablespoons of water by simmering together in a small pan. Add the butter and season.

Lay the chicory in a pan and pour over the liquor. Simmer for 30 minutes or until the chicory is cooked.

Remove the chicory. Increase the heat and reduce the liquor to a few tablespoons. Serve at once with the liquor poured over.

NON-FORCED CHICORIES

All such varieties stem from the wild chicory which was blanched and eaten by the Egyptians, Greeks and Romans, who all enjoyed the leaves in salads as well as cooked. There are now numerous varieties of green- and red-leaved chicories: radicchio, with its cardinal-purple leaves, is perhaps the most striking. With their colour and sharp taste, these make a welcome addition to winter salads.

VARIETIES To further confuse the shopper, in Italy all chicories are known as *radicchio*. They are often as beautiful in their colouring and shape as ornamental cabbage.

Rossa di Treviso has long pointed leaves and will flourish in British gardens. Its foliage is green in summer and turns red once the cold sets in. This variety is often sold young, when the root appears like a brownish radish with the leaves like a small mop of carmine red. It is worth buying it with its roots intact, for you can cut off the top and eat it as salad while burying the root in warm damp peat. Keep the container in the dark and in six weeks you should have your own red chicons.

Rossa di Verona forms a tight ball of leaves, thus protecting itself in the winter. Variegata di Castelfranco has green leaves blotched with red and forms a loose head of red and white leaves at its centre. Variegata di Chioggia has green leaves in the summer, but goes red and white in the winter months.

COOKING All the chicories above can be cooked as well as eaten raw. Simply treat them as you would lettuce, tearing off the leaves and cooking in butter and/or olive oil, or a little stock, or adding to soups and stews. However, cooking turns the red shades into green and I cannot see much point in it. The colour is retained, however, if the salad is served warm. All are slightly bitter in taste, but in some this is far more pronounced; they get less bitter the later in the season they are picked.

LEFT *Rossa di Treviso*

ENDIVE

Cichorium endivia

This group of salad plants is the one which is confused with chicories, understandably as they are close cousins and they were considered the same family in the Ancient World. They can be divided up into curly- and broad-leaved types. Like the chicories, they have always been used in salads and sometimes cooked. We also know the broad-leaved kind as batavia, escarole or scarole.

Curly-leaved types include Endivia Riccia Pancalieri, which has pronounced curly leaves with rose-tinted white midribs and a large white heart, and Wallonne, a French variety which forms a tightly packed head with a self-blanching heart.

Broad-leaved varieties include Scarola Verde, which forms a large head with green and white leaves, and Batavian Green, which is the most popular of all as it is vigorous and hardy.

WARM SALAD OF RADICCHIO AND GLAZED SQUASH

2–3 tbsp olive oil
2–5 garlic cloves
30 g / 1 oz root ginger, peeled and grated
1 small squash, peeled, deseeded and cubed
2 tbsp white wine vinegar
1 tbsp caster sugar
1 tsp sea salt
1 tbsp sesame oil
1 radicchio, separated and torn into pieces if necessary
½ curly endive, separated and torn into pieces
½ broad-leaved batavia, separated and torn into pieces
few coriander leaves, chopped

Heat the olive oil in a frying pan and add the garlic, ginger and cubed squash. Fry for about 5 minutes, turning the pieces so that they brown and are crisp on the outside.

Add the wine vinegar, sugar and salt, raise the heat and cook for another minute.

Meanwhile, heat the sesame oil in a pan and throw in the radicchio, endive and batavia leaves. Stir-fry for a minute or until the leaves just begin to wilt. Turn off the heat and transfer the leaves to a large serving platter.

Pour over the squash and sprinkle with coriander.

DANDELION

Taraxacum officinale

Our name for this plant is a corruption of the French *dent de lion*, from the medieval Latin *dens leonis*. The leaves, upright and pointed, are cut into jagged teeth and it is these which it is thought, rather fancifully, resembled the teeth of the lion. The Latin horticultural name is derived from the Greek *taraxos* (disorder) and *akos* (remedy), which shows how ancient were the curative and medicinal uses of the plant.

The first mention of dandelions as a medicine is in the works of the Arab physicians of the tenth and eleventh centuries. It is the milky white juice contained in the root which was valued as a stimulant and diuretic – it did not get its French name *pissenlit* for nothing. A brew of the roots has also been given in cases of hepatitis. Mrs Greaves tells us that: 'A broth of Dandelion roots, sliced and stewed in boiling water with some leaves of Sorrel and the yolk of an egg, taken daily for some months, has been known to cure seemingly intractable cases of chronic liver congestion.' As the roots are high in inulin (see page 76) one wonders whether the patient suffered torments of flatulence under this cure.

The dandelion is an important plant for bees and the production of honey, for the flowers open early in the year, even in a cool spring, and furnish considerable quantities of nectar and pollen, while successive blooms continue throughout the year until the late autumn. All this might make the gardener curse, especially those taking pride in their lawns, but the flowers also forecast the weather. They are highly sensitive and in fine weather the flowers follow the sun, but long before dusk the flowers close up against the dew of the night and will also close up at the threat of rain in the day.

DANDELION *Taraxacum officinale*

NUTRITION

The leaves are high in vitamins A and C and contain more iron than spinach, as well as potassium. They thus form a very valuable salad food, especially early in the spring when there are few fresh greens around.

PREPARING AND COOKING

Pick the leaves when they first appear, the small young leaves being by far the best. Older leaves tend to be bitter, but no more so than a cultivated endive for example. Use the leaves in mixed salads or on their own. Sandwiches of dandelion leaves, dressed in a little oil, lemon, sea salt and pepper with perhaps the addition of avocado or onion, are excellent.

The leaves can also be cooked and what better way to use the dandelions in your garden than to pick all the leaves and sauté them in a little butter and oil over a gentle heat for about 10 minutes?

GLOBE ARTICHOKE

Cynara scolymus

Both the Greeks and the Romans were very fond of artichokes, or was it the cardoon? We are uncertain, for the globe artichoke is the cultivated form of cardoon. It is thought that these handsome plants were discovered in North Africa and then cultivated in southern Europe.

However, it was not until the middle of the fifteenth century that the artichoke was cultivated for its huge unopened flower. In Venice in 1493, the chronicler and historian Ermolao Barbaro knew of only one single plant grown as a novelty in a private garden, but it soon became a staple and popular food all over Italy. Not to Goethe, however; in the eighteenth century in his *Travels Through Italy* he comments that the peasants ate thistles and that he found them not to his liking. These could actually have been thistles, for the Romans ate them and it is reasonable to suppose that the Italian peasants continued to eat them when fashionable society had long forgotten about them. The artichoke did not reach England until the middle of the fifteenth century, but became popular in the kitchen garden. The plant grows well in our climate and by 1687 it was thought by John Worlidge – who wrote the first systematic treatise (*Systema Agriculturae*) on agricultural husbandry on a comprehensive scale – to be 'one of the most excellent fruits'.

Artichokes have a long history of medicinal use; indeed the Romans thought the juice was a cure for baldness. Dodoens' *Herbal* of 1578 thinks the root is good against the rank smell of armpits. It notes also 'that it sendeth forth plentie of stinking urine' but added that this has the positive effect of amending the 'ranke and rammish savour of the body'.

John Evelyn tells us that in Italy artichokes were eaten with orange juice and sugar. He himself, surprisingly, cuts these small heads into four and eats them raw with vinegar, salt and pepper, washed down with a glass of wine – following Dr Muffet's advice – at the end of the meal. Evelyn also cooks the young artichokes in butter until crisp and serves them with parsley, while he bakes the bottoms, or *fonds* in French, in pies with beef marrow and dates.

I grow more artichoke plants every year, as we harvest the flower buds when they are small. These only need their pointed leaves cut away, then the outside of the bottom peeled and the merest few outside leaves discarded. The hairy choke has not formed by that stage, so those inner yellow leaves can just be gouged out with a knife. Everyone now tells you to place the bulb in acidulated water until ready for cooking to stop discolouring, but it is only a matter of minutes before a dozen or so small artichokes are trimmed and prepared. These now need to be quartered, then fried in olive oil with some garlic. They are then served with a sprinkling of sea salt and lemon juice. Cooked this way they are the most delicious of all foods that are garnered from our garden. Baby artichokes can also be trimmed and peeled, then sliced thinly downwards, dipped in a light batter and fried until crisp.

However, for many people eating the artichoke has a curious effect on their taste buds. American food writer James Beard commented in 1971 that wine drinkers felt that artichokes ruined the flavour of fine wines. Research was conducted in the '30s

and it was discovered that for two-thirds of people eating artichokes made a drink of water seem sweet. In 1954 it was discovered that the source of this phenomenon was an organic acid called cynarin, unique to artichokes. This substance appears to stimulate the sweetness receptors, but only in those people who are susceptible to it.

VARIETIES

There are two main varieties of artichoke, the green and the purple. The latter has rather spiky leaves and can be more awkward to prepare, but there is no difference in flavour. Artichokes also come in round globes or egg shapes; again there is no noticeable difference in flavour. There are dozens of different types and each country seems to have a preference.

BUYING AND STORING

Globe artichokes go brown and dry out on the outside very quickly (inside a week), so pick your globes when the exterior leaves are green and juicy. If you are not going to eat them at once, keep them in the salad drawer of the refrigerator for not longer then three days. They will keep just as well if you cook them and then refrigerate for two days. All vegetables, however, taste better if cooked and eaten as soon as possible after being bought.

PREPARING AND COOKING

The habit of eating the artichokes when young is common all over southern Europe, in Italy, Spain and Greece. Sadly, however, young artichokes do not get exported. The large artichoke season is short, for most of our artichokes come from Brittany, where they do not harvest them until they are as plump and large as an ogen melon. I once asked that they should export the artichokes when small, starting in March when they could continue through until July, but my request fell on deaf ears. Brittany artichokes had to be huge and plump and that was the way the ignorant

GLOBE ARTICHOKE *Cynara scolymus*

British would continue to get them. That, at least, seemed to be the sub-text of their indifference.

If you are serving them whole, these large artichokes need to be boiled for anything from 30 to 45 minutes. Pull a leaf out after half an hour to see how tender its base is. Large globe artichokes need to have their bases trimmed so that they sit steadily on the plate; they can also have their tops severed. Restaurants often serve them in this way, but you will need a sharp and heavy knife for this operation. When you need the flavour of the artichoke in the stock for a stew you need to perform this rather radical trimming and preparation on the raw vegetable. I use a cleaver, which cuts the top half of the leaves away. Now, before it is cooked, the choke can be scooped out with a sharp knife and then just the bottom ringed with leaves can be boiled, ready to be used as a receptacle for any flavourings.

For certain of the recipes below I think this kind of surgery is worth doing on the large artichokes. This raw receptacle or cup can then be quartered and used in various stews. For other recipes, however, this process can be done later, after the artichoke is cooked. These recipes are the starters, when the artichoke is eaten just with a dressing or served with a simple stuffing. If you decide to keep the leaves on, the most attractive way of presenting the artichoke is to fan out the exterior leaves so that it looks like a water lily, then pluck out the purplish core of inner leaves and cut out the choke beneath that and serve on individual plates. On the other hand, you can choose to serve the artichokes as they are after being cooked and let the individual diners perform the surgery. Do then provide a large platter in the centre of the table for all the discarded leaves.

One can occasionally find artichoke plates with a recess in the centre for the sauce, surrounded by leaf-like indentations – pretty, but not very practical. Small individual ramekins filled with sauce can be supplied in which each diner may dip their leaves. The next question is what sauce to serve with the cooked artichokes?

If the artichokes are served cold, a garlicky vinaigrette is the most common sauce; if warm, then a melted lemon butter is probably the most satisfactory. Enormous pleasure can be achieved by experimenting with other sauces for the artichokes, whether they are served hot, warm or cold. Try buttermilk, smetana or yogurt flavoured with lemon, garlic, chilli, honey, or soy sauce and celery salt perhaps, even mustard, oil and basil. You could go to town with a simple Hollandaise or *aïoli* mayonnaise, or a mixture of a not-too-hot rouille whipped into yogurt and olive oil.

If you choose to present the artichoke stuffed, then the choke and inner leaves have to be plucked and cut out, so that the delicious bottom is exposed. All manner of flavoursome fillings can then go into the centre, but they must be fairly moist as the outer leaves need to be dipped into the filling instead of a sauce. At the same time, the filling must be stiff enough to stay where it is. Purées of various kinds are the answer. A classic is the Broad Bean Purée (see page 154), flavoured faintly with savory, and a purée of fresh garden peas is also marvellous, as is a purée made from leek, mushroom and courgette. A rather popular starter in restaurants can be artichokes filled with prawns in some dressing. This seems to me to be gilding the lily. If desired, prawns in a dressing can be eaten separately on a bed of leaves.

A favourite Victorian recipe, which needed a mountain of labour for what one imagines was a mere mouse of result, was *ARTICHAUTS À LA BARIGOULE*. This consisted of 'parboiling them, taking away the choke, stuffing them with a mixture of onion, mushroom and parsley and adding a little chopped bacon. Then you tie the artichoke up with a cotton, wrap fat bacon around it and braise it. When done, take away the bacon and serve.'

You can now buy packets of frozen artichoke bottoms and these make excellent starters used in the way that the kohlrabi base is used (see page 105). Or if you are feeling reckless, these artichoke bottoms can be used in other ways. For example, they are delicious in pies and vegetable stews. An eighteenth-century recipe of Hannah Glasse's uses layers of artichoke bottoms with butter, egg yolks, mace, morels and truffles, moistened with a little stock and white wine and then baked in puff pastry.

A simpler recipe is one of Mrs Leyel's, where the artichoke bottom is served with a poached egg on top, covered with artichoke paste from the leaves and sprinkled with Parmesan. As one can now buy artichoke paste in bottles, this is a fairly trouble-free recipe and enormously pleasant. Uncooked egg yolk mixed with artichoke and Parmesan is unusually good.

ARTICHAUTS À LA PROVENÇALE

This recipe for small artichokes, another of Mrs Leyel's, is a complete winner.

Take young and tender artichokes, remove the outside leaves, trim the others with scissors and rub each artichoke with lemon to prevent it blackening. Sprinkle them one by one with good olive oil and put sea salt and pepper between their leaves.

Put them into a saucepan with some olive oil (enough to soak the artichokes) and then pour in enough cold water to cover. Put the saucepan on a very high heat so that the oil and water may boil quickly. In 20 minutes the artichokes will be cooked. Serve the artichokes on a very hot dish, pouring over the boiling oil which is left. They should be eaten off very hot plates. The whole of the little artichokes are edible when they are cooked in this way. Fresh water should be drunk after eating them; wine will seem acid, while water will have a delicate sweet taste.

POACHED ARTICHOKES IN WHITE WINE

4 artichokes (with stems if possible)
3 tbsp olive oil
3–4 garlic cloves, sliced
200 ml / 7 fl oz dry white wine
sea salt and freshly ground black pepper
julienne strips of carrot to garnish (optional)

First prepare the artichokes: take off the outside leaves to expose the lighter green interior leaves, then cut the top two-thirds of the artichoke away. Slice the bottom third in two and, with a sharp knife, cut out the choke and the middle cluster of leaves.

Now heat the olive oil in a pan and add the garlic. Place the artichoke halves in the oil, cut side down. Fry for a minute or two, then add the wine and seasoning. Cover and poach for about 10 minutes.

This can be eaten hot or cold, garnished with carrot if you wish. Mop up the juices with bread.

Note: if the artichokes have stems, these can be peeled and chopped into 2.5 cm / 1 in lengths, then cooked with the rest of the artichoke. They taste like the artichoke bottoms, but perhaps a little sweeter.

GRATIN OF ARTICHOKE AND ASPARAGUS

4–6 globe artichokes, prepared, trimmed and quartered
450 g / 1 lb asparagus, trimmed
2–3 tbsp olive oil
3–4 garlic cloves, thinly sliced
3 slices of wholemeal bread, cubed
4–6 small shallots, left whole
2–3 courgettes, sliced thinly lengthwise
little chopped parsley or lovage
sea salt and freshly ground black pepper

Throw the pieces of artichoke into a pan of boiling salted water and simmer for 20 minutes. Steam the asparagus for 10 minutes. Drain both vegetables well and slice the asparagus into 5 cm / 2 in lengths.

Heat the olive oil in a large pan and throw in the garlic, cubed bread, shallots and courgettes. Stir-fry for as long as it takes to crisp the bread.

Add the artichokes and asparagus. Stir-fry for another 10 seconds, season and sprinkle with the herbs.

SPRING ARTICHOKE CASSEROLE

3 tbsp olive oil
6 garlic cloves, sliced
4 or 6 large artichokes, trimmed down to their bottoms
450 g / 1 lb small new potatoes, scrubbed
450 g / 1 lb young broad beans, podded
450 g / 1 lb French beans, trimmed
300 ml / ½ pt vegetable stock
300 ml / ½ pt dry white wine
1 tsp beurre manié (mixture of equal parts flour and softened butter)
sea salt and freshly ground black pepper
handful of finely chopped parsley, to garnish

Heat the olive oil in a large saucepan and throw in the garlic and trimmed artichokes. Let cook a little and then add the rest of the vegetables, together with the stock and white wine. Simmer for 20 minutes. If it gets too dry at any point, add a little more of the stock or the white wine.

Increase the heat and thicken with the beurre manié. Season, then serve sprinkled with the parsley.

VARIATIONS

This works well because the artichoke gives something of itself to all the other vegetables, but variations to this recipe also work perfectly a little later in the season. For example, you can use peas, leeks and courgettes instead of the two kinds of beans. I have also used haricot beans and, best of all, flageolet with fresh garden peas, but replacing half the oil with 30 g / 1 oz butter and halving the garlic.

HOT STUFFED ARTICHOKES

4 large artichokes, trimmed
for the stuffing:
1 small tin of anchovies
115 g / 4 oz canned tomatoes
handful of finely chopped parsley
1 tbsp breadcrumbs
1 tbsp capers
sea salt and freshly ground black pepper

Boil the trimmed artichokes for 30 minutes. Take them out of the pan and open out the leaves. Extract the central cone of leaves and dig out the choke beneath to expose the bottom. Arrange the prepared artichokes in a baking pan.

Meanwhile, preheat the oven to 190°C/375°F/gas5 and make the stuffing: cook the anchovies and tomatoes with their liquid together, stirring so that they become a thickish purée. Add the parsley, breadcrumbs and capers to thicken and flavour the purée more. Season and spoon this into the centre of the artichokes.

Warm through in the oven for 10 minutes, then serve at once.

CARDOON

Cynara cardunculus

You may be fortunate enough to eat this vegetable in Spain and would be most likely to find cardoon in the restaurants of southern Spain. Its season is in late autumn and throughout the winter months. The stalks are blanched like celery and they are cooked by being poached in stock. They are highly delicious, having a flavour somewhere between asparagus and artichoke. No wonder the Ancient World was so enthusiastic, Pliny said cardoon was so esteemed in Rome that it fetched a higher price than any other garden herb.

Once you have eaten cardoon you may, like myself, be unable to control the desire to have a steady supply and decide to grow them in your own garden. They grow easily in Britain and are a handsome plant, tall, stately and silver, with rather smaller thistle heads than the globe artichoke. Like artichokes, the flowers of cardoon have been used throughout history as a source of vegetable rennet.

What is not easy is to decide what method of blanching to use to get the cardoon to the table and what time of year to do it. Two methods appear to have been used. As cardoons are ruined by frost and an early hard winter will destroy them, the stalks were wrapped in brown paper in August and blanched for several months before being lifted in October. (The whole plant in its first year would be dug up by Victorian gardeners and the bulb eaten as well as the stalks.) The other method, which I adopt, is to blanch the new shoots as they appear in January and to keep them warm by covering the cloches with sacking, much as sea kale is cultivated. Both methods have been used from ancient times. Cardoons are perennial, however, so a long-lived plant will not be the best to blanch with brown paper in the autumn as the stems will have become too fibrous.

Cardoons have never been grown with much enthusiasm in northern Europe and Britain, except as an unusual plant for the herbaceous border. Yet John Tradescant, the traveller, naturalist and gardener, saw several acres being cultivated in Brussels in 1629, but had no idea how it was harvested or cooked.

Some authorities inform us that the flower heads can be eaten in the same way as globe artichokes. They can, but in my view they are not worth the trouble. The base of the flower (the equivalent of the artichoke *fond*) has neither the flavour nor the tenderness after cooking found in the artichoke.

PREPARING AND COOKING

Hannah Glasse sums it all up succinctly: 'You must cut them about 10 inches, and string them, and tye them up in Bundles like Asparagus, or cut them in small Dice, and boil them like Peas, and toss them up with Pepper, Salt and melted Butter'.

Young small plants can be eaten raw. In the markets of southern Spain, Italy and France you may well see cardoon hearts laid out on sale tied together. Discard the outer stalks, which are too fibrous, and trim the rest, putting the pieces into acidulated water so that they do not discolour. Cut into bite-size pieces and dress with vinaigrette or a light *aïoli* mayonnaise.

The same stalks can be cut to the length of asparagus and poached in stock for 10 minutes. In Italy they are often eaten raw with *bagna cauda* ('hot bath'), that delicious sauce of anchovy and garlic. The cooked cardoon is also traditionally served with *ANCHOVY SAUCE*. This is simplicity itself to make. Heat 150 ml / ¼ pt of olive oil with the contents of a small tin of anchovies and 2 or 3 garlic cloves for a few minutes, until both the garlic and anchovies have dissolved.

As cardoons are eaten in southern Europe in autumn, they are often teamed with nuts. Cardoons can be poached in stock with a few walnuts (even a little chopped bacon if you feel like it) and then served simply dressed with walnut oil and lemon. Alternatively, poach them with chestnuts in stock, with the addition of a little white wine. The sauce is then blended to make a purée and poured over the cooked cardoons.

CARDOON *Cynara cardunculus*

JERUSALEM ARTICHOKE

Helianthus tuberosus

This vegetable's name is said to stem from the Italian *girasol articocco*, 'sunflower artichoke'. Soon after it was brought back from the New World it was called the 'Canada potato' and 'French potato', for it flourishes in the central United States and Canada and, being somewhat of a pest in its spreading habits, had easily reached the Atlantic coast long before the white man did. Discussing its countries of origin in 1633, the revised edition of Gerard's *Herball* was right in mentioning Canada, but wrong when he added Peru and Brazil. Gerard also notes 'that this plant hath no similarity in leafe, stalke, root, or manner of growing, with an Artichoke, but only a little likenesse of taste in the dressed root.' Another early observer, Marc Les Carbot, said 'Most excellent to eat, tasting like chards or cardoons, but more pleasant...'

It was an English gardener, John Goodyer, who was given some tubers by a Flemish merchant in 1617 and first grew the Jerusalem artichoke in Britain. It was also he who first noted the vegetable's tendency to bequeath flatulence. Goodyer writes '...they stir up and cause a filthie loathesome stinking winde within the bodie, thereby causing the belly to bee much pained and tormented, and are a meat more fit for swine, than men.' Harold McGee in *The Curious Cook* has written amusingly and authoritatively on the reasons for this vegetable's flatulent ability in a chapter called 'Taking the Wind out of the Sunroot'.

The Jerusalem artichoke contains no starch or oil and little protein, 50 per cent of it is carbohydrate which is indigestible to us. McGee says 'the troublesome molecules have never been a sufficiently important part of our diet for us to have evolved enzymes to digest them.' The artichoke lays down its reserves of energy for next season's sprouts in fructose chains named inulin, and it is this chemical which gives trouble. The curious fact is that these tubers do not affect everybody in the same way. Some fortunate people can tolerate the vegetable 'by virtue of native intestinal flora and digestive physiology'.

This was true of myself, for I ate Jerusalem artichokes with enthusiasm up to my mid-forties before they turned on me. McGee suggests methods that go some way to tame the vegetable (see below). I am saddened by the effect of this phenomenon, for I used to find the flavour delicate and delicious, while a soup made from the tubers was always warmly received at dinner parties.

VARIETIES AND GROWING

Twenty years ago Jerusalem artichokes were beige-coloured and very knobbly and a great deal of bother to clean. Since then varieties have got smoother and tend to look now very much like a delicious type of potato – the Pink Fir Apple. Boston Red has an attractive rose-red skin and a few knobs on it. Fuseau is long and white, with tubers that tend to be smooth.

If you want to grow Jerusalem artichokes, tubers can be sown from the ones available at the greengrocer. Only grow them if you have plenty of land, they will grow 2 m / 6 ft high and spread alarmingly. Like the potato, they do have the advantage that they can be left in the ground until you need to dig them up. However, the smallest sliver of tuber left in the ground will send out shoots in the spring, and once planted they are difficult to eradicate.

STORING, PREPARING AND COOKING

The tubers store well in a dark cool place. They need washing under cold running water, and perhaps scrubbing if the dirt is obstinate – do not peel them.

McGee tells us that the flatulent effect can be controlled by the length of time the artichokes are cooked. Even a modest 10–15 minutes' boiling in copious amounts of water will extract some of the inulin. To maximize this effect, the artichokes can be cut into slices, which increases the surface area that is exposed. McGee estimated that 15 minutes' cooking drew out almost half of the inulin. He also discovered that temperatures just above freezing affect the vegetable so that it pre-digests its own carbohydrates, thus breaking down the inulin. So artichokes bought in March and April will be more digestible than those bought before Christmas.

Like parsnips or potatoes, artichokes will roast happily after an initial boiling, but artichokes can be boiled longer than the requisite 15 minutes before being drained and added to the roasting pan. After boiling, they may also be sliced and fried in oil with garlic, or dipped into a light batter and fried. I now view the soup as a rather dangerous choice, for the inulin that is leached out of the vegetable is the rather cloudy, slightly sweet precipitate in the liquor that the cook needs to make a flavourful soup. For those readers untouched by the explosive charge of this vegetable, I give the recipe.

JERUSALEM (OR PALESTINE) SOUP

for 6

For years throughout the eighteenth and nineteenth centuries, the soup was commonly referred to as Palestine or Jerusalem soup and only lost these names in this century.

Mrs Beeton's recipe begins with 3 slices of lean bacon or ham, half a head of celery and both a turnip and an onion before adding 1.8 kg / 4 lb of artichokes.

She also suggests that a pretty way of serving Jerusalem artichokes is to shape them like pears, then cover these in white sauce and finish with a garnish of Brussels sprouts.

30 g / 1 oz butter
2 tbsp olive oil
1 onion
675 g / 1½ lb Jerusalem artichokes, cleaned and chopped into dice
1.5 litres / 2½ pt vegetable stock
575 ml / 1 pt milk
sea salt and freshly ground black pepper
little chopped parsley or coriander, to garnish

Melt the butter with the olive oil in a large saucepan. Add the onion and the artichokes. Cook for a moment, stirring the vegetables, then add the stock and seasoning.

Let the soup simmer for 30–40 minutes, then add the milk and leave to cool.

Blend in a liquidizer and then reheat carefully. Before serving, sprinkle a little chopped parsley or coriander over the surface.

ABOVE Salsify

SALSIFY and SCORZONERA

Tragopogon porrifolius and *Scorzonera hispanica*

These two long thin roots belong to the same family and taste very similar. Both are white-fleshed, but scorzonera has a black skin. When young, their leaves can be eaten in salads, but they are both grown for their roots. Like those of the Jerusalem artichoke and dandelion, these are high in inulin, but their flavours are so delicious that they are well worth eating.

Salsify is also called the 'oyster plant' as its taste was thought to resemble oysters – with a stretch of the imagination, perhaps. Others disagree: Dorothy Hartley comments, 'and so like oyster that it takes in the cat!' The taste of both is really like that of a subtly flavoured globe artichoke with a hint of asparagus.

The Latin term *tragopogon* stems from two Greek

words meaning 'goat's beard', thought to describe the fluffy character of the seed ball. The fact that the flower head closes at noon, not opening until the following morning, gave it its other popular name, 'Jack-go-to-bed-at-noon'. Salsify is often called purple goat's beard, its flowers being purple rather than the yellow of *Tragopogon pratensis,* a type eaten in the seventeenth century but now forgotten. The name salsify is a corruption of the old Latin *solsequium*, meaning 'sun' and 'following'.

The whole plant was used medicinally: Culpeper advises using roots, flowers, leaves and all bruised and boiled and then strained, as a remedy for heartburn, loss of appetite and disorders of the breast and liver. It was not cultivated much before the seventeenth century. One of the first to mention it is John Evelyn, who refers to it as viper-grass. After pointing out its medicinal virtues, Evelyn goes on to say how good it is stewed with beef marrow, spice and wine. He also says how pleasant it can be raw in a salad.

Evelyn called scorzonera 'viper-grass' because the soup was considered an antidote to the poison of the viper. Its other names are black salsify, black oyster plant and even black radish. It is a member of a different genus altogether. It came from Spain and crept into the rest of Europe, especially France, Belgium and England, whose herbalists and gardeners all took an interest in it at the end of the sixteenth century. The astonishing fact is that the flavour is hardly distinguishable from salsify and, once peeled and cooked, I would defy anyone to tell the difference.

PREPARING

Wash both roots under cold running water, without peeling, and boil for 10 minutes. When cool, the skin will come off like a glove, especially if placed again under cold running water. The roots tend to be thin and it is easy to waste the vegetable if you attempt to peel them beforehand. Besides that there is then all the chore of plunging the peeled roots into acidulated water to prevent discoloration.

COOKING

After peeling the cooled cooked roots, the easiest and nicest way to eat both is as a salad, tossed in a garlicky vinaigrette and served with other salads. If the vegetable is to be eaten hot, however, there are numerous methods: toss in butter and sprinkle with parsley or chives; dip the sliced roots in a light batter and fry; or stir-fry with ginger, garlic and chilli.

Polygonaceae

RHUBARB

Rheum rhaponticum

I am including rhubarb in this vegetable book, as I tend to use it as such. It can make excellent soups, pickles and sauces for fish. Besides, I am fond of its bright astringency and my childhood spirits were never clouded by such a dish as stewed rhubarb and custard. Though it is famous as the ingredient in sweet tarts and pies, I believe its use and enjoyment should be widened far away from the dessert course.

The first mention of rhubarb is for its medicinal use in China. This was *Rheum officinale,* which flourished in north-west China and was favoured in the Pen-King herbal from 2700 BC. It is believed originally to have grown somewhere in northern Asia – Mongolia, Tartary, or perhaps Siberia.

It reached the Mediterranean some time around the birth of Christ, for both Pliny and Dioscorides mention it, though both rather dismissively. China began exporting rhubarb – the dried roots, that is – to the West very early on and it began to collect various synonyms. There was Russian rhubarb, Turkish and Chinese, all named after the various trade routes. Prior to 1842, Canton was the only port through which the Chinese empire traded, so rhubarb had to travel overland to Russia via the frontier town of Kiachta, to Turkey through the Levant. It got its name *Rha* from the ancient name of the Volga, on whose banks the plant grew. The Volga flows into the Black Sea – 'the wine dark sea' of Homer. *Pontus* is Greek for sea, so *Rhaponticum* indicates its trade source.

The root was used in the West as a curative potion throughout the Middle Ages. It was also growing in monastery gardens to stock abbey pharmacies, but it does not seem to have occurred to anyone to eat it.

When they did, they began by eating either the root or the leaves and might have felt unwell enough not to try it again. The leaves were thought to be toxic, containing oxalates of both potassium and calcium. It is, then, somewhat surprising that Green says in his *Universal Herbal* (1832): 'The leaves are also used by the French in their soups, to which they impart an agreeable acidity like that of Sorrel.' He was not alone in this view; the gardener of the Earl of Shrewsbury at Alton Towers said he used rhubarb flowers before they opened in the same way as broccoli, cooked au gratin with white sauce; the cheese, it was thought, might have obscured any bitterness.

However, rhubarb stalks, which we eat happily, also contain some degree of these oxalates. Also, when poisoning occurred it affected some but not others, or could some people be more susceptible to oxalate poisoning? It is doubtful. McGee tells us that rhubarb has had a somewhat shady reputation in the USA since World War I when Americans were encouraged to eat the leaves as a vegetable supplement and many were poisoned as a result. Now it seems that the toxins in the leaves are not the oxalates, but something else which remains to be identified.

In western Europe by the sixteenth century, rhubarb was thought to be a cure for venereal disease. Two ounces of dried root and half an ounce of parsley were boiled in two quarts of water, then reduced by two-thirds. This was drunk several times a day and must have been punishment enough, without being saddled with the affliction.

As a vegetable/fruit, it was one of the last to be cultivated in Britain. It continued to flourish in Tudor herb gardens and, as late as the 1770s, the Duke of Atholl had a plantation of rhubarb on his estate at Blair Castle in Perthshire, which was all sold to an Edinburgh druggist. By that time it was beginning to creep into English cooking.

The apothecary to James I, John Parkinson (1567–1650), first planted rhubarb from Italy in his own garden, but it took about 150 years before it was accepted as a fruit for tarts. A recipe of 1790 gave directions for slicing the stalks and then treating them like gooseberries. By the time of Mrs Beeton there are recipes for tarts and boiled puddings with rhubarb, as well as some others for wine and jam.

There, in these dishes, rhubarb found its niche. No kitchen garden was without its clump, often forced in early spring by digging out some of the roots and leaving them for the 'frost to get at them'. The roots were then placed in the dark, and in a few weeks the shoots began to grow. A lively nacreous pink, for their colour alone they were worthy of any table, but as their acid content was high they were cooked with an excess of sugar. Because of this, those dishes have been unpopular lately. Yet country gardens still grow rhubarb and it still features in jams and tarts, and was once famous for the wine. Indeed, Mrs Leyel gives a recipe which uses 5 lb of rhubarb to 3 lb of sugar and a gallon of water.

Rhubarb also makes an unusual but excellent soup. However, it only works if the stock used has a strong savoury flavour, for this must offset the acid in the rhubarb; otherwise it needs a quantity of sugar and you will have a cordial for dessert and not a savoury soup. The most successful version of this soup I made was when I used the carcass bones from smoked quail for the stock. However, any smoked game would do: pheasant, mallard or guinea fowl. I imagine, though I have not tried it, the carcass of smoked chicken or a ham bone would be excellent. The method is simplicity itself, and the soup can be served either hot or cold. Its flavour is often difficult for guests to pin down, while the colour is jewel-like.

RHUBARB SOUP

900 g / 2 lb rhubarb, trimmed and chopped
2.25 litres / 2 qt strong savoury stock, strained
1 or 2 tbsp sugar
Tabasco
sea salt and freshly ground black pepper
2–3 tbsp sour cream, smetana or yogurt
few geranium petals for garnish

Cook the rhubarb in stock for 10 minutes or so, cool and blend to a thin purée. Make sure it is liquidized thoroughly, as strands of fibrous rhubarb floating in the soup look unsightly and give the game away.

Taste with care: depending on how strong the

ABOVE *Pickled rhubarb*

stock is and how salty, it may need a spoonful of sugar, or even a few drops of Tabasco, for the soup has to be a balance between saltiness and astringency.

Spoon some sour cream, smetana or yogurt and float a few geranium petals on top before serving.

RHUBARB SAUCE FOR FISH

30 g / 1 oz butter
450 g / 1 lb rhubarb, trimmed and chopped
½ tsp sea salt
½ tsp caster sugar
¼ tsp Tabasco

Melt the butter and add the remaining ingredients. Simmer for 10 minutes over a low heat, leave to cool.

Blend in a liquidizer. Taste and adjust the seasonings, then refrigerate until needed (an ice-cold sauce tastes particularly good with grilled white fish).

PICKLED RHUBARB

This makes an excellent pickle for game, meat or cheese. It also looks charming on the plate. The rhubarb is 'cold cooked' in the *ceviche* method the South Americans use for fish.

900 g – 1.35 kg / 2 – 3 lb rhubarb, trimmed and chopped
30 g / 1 oz root ginger, peeled and thinly sliced
1 tbsp cloves
575 ml / 1 pt or more cider vinegar
1 tbsp sea salt
2 tbsp caster sugar

Pack the rhubarb into a sterile preserving jar, interspersed with a few slices of ginger and a few cloves.

Heat the vinegar and add the salt and sugar. Pour over the rhubarb and place an airtight lid on the jar.

The pickle is ready in three weeks.

SORREL

Rumex acetosa

This is a beautiful vegetable, with a flavour of gooseberry unique among green vegetables, that is invaluable in the kitchen. It is astonishing that it is not commercially grown and readily available, but we can easily grow the plants in our own kitchen gardens. Even if you have a very small plot indeed, a root of sorrel will be worth it. The leaves spring up early in the year and can be picked for salads when there is little else that is green around. Later, soups and sauces can be made from the leaves. The plant will also continue to flourish late into the season and throughout the first frosts.

Sorrel was used and prized in antiquity. Of all vegetables, sorrel has the one of the largest roots, Pliny thought, 'going as far as a yard and a half into the ground and its root is full of sap and lives a long time, even after being dug up'. This is quite true; some sorrel roots I found mistakenly discarded on a compost heap two weeks after being thrown out were dug back into the soil and were found to be astonishingly healthy the following spring.

In ancient Rome the acidity of sorrel induced cooks to consider that it possibly broke down the tough tissues of meat or the bones in fish. A classic dish today is *alose à l'oseille* (shad with sorrel), where the fish is cooked on a bed of sorrel in the belief that the small bones in the fish will melt. A myth, of course, and they have to be filleted out. The shad is one of the great fishes of the world, with a delicately meaty flavour which only needs brief cooking, and a sorrel purée is indeed a perfect complement.

The whole family of the Polygonaceae is known for its astringency, caused by the presence of tannins which are found mainly in the roots. Another member, bistort (*Polygonum bistorta*), was used in the spring for its leaves in the north of England as an ingredient in herb pudding. Its root, once it has been steeped in water so that the tannin leaches out, can provide a flour. In hard times in Russia, Iceland and Siberia, such a flour was used to make bread. Bistort used to be widely cultivated and that which grows wild in Britain now is the progeny of those that escaped from gardens long past. In medieval times its uses were legion, for all bowel complaints and in haemorrhages from lungs and stomach, or as a remedy for nose-bleeding, for it is one of the most astringent medicines in the vegetable kingdom and is highly styptic. A distilled water of the leaves and root was used to soothe stings and bites, and also as a throat gargle.

Culpeper thought a syrup made from sorrel itself, if gargled in the mouth, would cure sores. Gerard says that the root of bistort is good for looseness of teeth and hardening of gums, 'being holden in the mouth for a certain space and at sundry times.' He also goes on to say of sorrel: 'The juice hereof in summer time is a profitable sauce in many meates and pleasant to the taste. It cooleth a hot stomacke; mooveth appetite to meate; tempereth the heat of the liver, and openeth the stoppings thereof.'

In the time of Henry VIII, sorrel had an important place in all the kitchen gardens. William Turner referred to it in 1538 as both 'sorell' and 'sourdoc'. Yet 150 years later, in 1699, the zoologist Dr Lister was surprised to discover how much sorrel was cultivated around Paris. 'There is so much taste for sorrel here that I have seen whole acres devoted to it ... nothing is healthier ... it could be a substitute for the lemon in the treatment of scurvy and its related afflictions'.

From the earliest times, sorrel was cooked with cream, butter and eggs. Indeed, this fusion is highly satisfying; the richness of the cream is refreshing, spiced with a sharp, acidic, fruity, gooseberry flavour. A sauce was also made by cooking the leaves and mixing them with sugar and vinegar to accompany roast pork or goose. Hannah Glasse gives a recipe for poached eggs served on a bed of sorrel with plenty of brown three-cornered toast stuck around the plate, served with quartered orange. The juice was even used for taking rust marks out of linen.

On the whole, however, the English tended to disapprove of sorrel: a vegetable so acid surely could not be good for you? Besides, it contained oxalic acid and too much of that was poisonous. Disbelievingly, E. S. Dallas comments: 'But the French will eat it as a dish by itself; taking a whole peck of leaves to make a purée to go with a fricandeau of veal or with poached eggs. We all like some acid to go with our veal, and sorrel is a favourite accompaniment of certain fish; but after all there is no acid comparable to lemon juice for delicacy of flavour and for wholesomeness. Let us reserve sorrel for Bonne Femme Soup and wood-sorrel for Julienne.'

We now, of course, think of *potage à la bonne femme* being made with leeks, though André Simon gives a recipe for it with onion, lettuce and sorrel. The name paints a picture of a good woman and more or less means now a soup in the simple or rustic manner. Possibly the goodness comes from the industry and skill of a good cook who can contrive a soup made of whatever few vegetables are at hand. But what is the term 'julienne' doing here? We think of it as being a description of vegetables cut into tiny straws. Dallas gives a fascinating interpretation of how match-stick vegetables came to be called thus and it all rests on a tiny wild herb.

Wood sorrel (*Oxalis acetosella*) is not part of the Polygonaceae at all, but in flavour the leaf is a delicate version of *Rumex acetosa* and since antiquity has been used to flavour soups and salads. Do grow a plant in your own garden; they are not only delightful in their flowers with five white purple-veined petals, but the trefoil leaves are also charming.

The plant has a host of common names, including alleluiah and hallelujah. In Somerset it is called 'bread-and-cheese-and-cider', in Cornwall 'cuckoo's bread', in Dorset 'sleeping clover' as the leaves fold back. Why on earth 'hallelujah'? Turner in his *Herbal* of 1568 explains because it 'appeareth about Easter when Alleluja is song agayn, it is an expression of Christian celebration, an image of Christ risen'. Another Christian explanation is that the wood sorrel is the true shamrock and that when St Patrick beheld this emblem of the Trinity he praised God.

But might it also have been a pagan cry of delight for the return of spring? For the name hallelujah for wood sorrel appears in French, Spanish and Italian too. Dallas suggests that allelujah among peasants was corrupted into 'Lujula' and in Italy this became 'Julida' – or 'little Julia'. In France, in turn, this became 'Julienne'. (My French dictionary, Harrap, dated 1957 incidentally says of Julienne, '(a) Bot: Rocket (b) julienne soup'. Thus implying that the soup would be made from rocket.)

But back to Dallas and his ingenious explanation. Wood sorrel is a small, low-lying plant in which the trefoil leaves grow at the end of slender, thread-like stalks. To make a soup from wood sorrel, a bunch of leaves would be poached in stock, the leaves disappearing (as is the wont of all types of sorrel leaves) while the fine straws of the stalks remain. What we then have is a tasty stock with slender stalks floating in it. So when cooks made the soup from common sorrel, the leaves flavoured the stock, but there were no stalks, so cooks put threads of carrot, turnip and celery into the julienne to simulate them.

Now, of course, nobody makes julienne soup. The end result would hardly equal the labour involved, all that detailed chopping of vegetables. Larousse, it is interesting to note, gives a recipe for julienne soup, but without sorrel – wood or otherwise.

There have always been varieties of sorrel other than *Rumex acetosa*; sometimes one is not certain

BELOW *Sorrel growing*

which variety past herbals are referring to. The main ones are French sorrel (*R. scutatus*) and sheep's sorrel (*R. acetosella*). The first, as you would expect, is the type favoured by the French in cooking. There is, in my opinion, no difference in flavour once cooked, though the plant's leaves are buckler-shaped and almost brittle. Sheep's sorrel is a much smaller plant, its leaves tinged with red at the end of summer. Both this and garden sorrel are indigenous to the British Isles and can be found in pasture and fields.

Yet another member of the family was also used as a staple food plant by American Indians. Solomon's seal (*Polygonatum multiflorum*), which we grow in the herbaceous border for its elegant arching stalks and bell-like pendulous flowers, has its tubers and roots near the surface. These have been eaten for their starch. They were first of all soaked in water to leach out the tannins, then dried and pounded. The young shoots of Solomon's seal were also used in Turkey, where they were boiled and eaten like asparagus.

PREPARING AND COOKING

John Evelyn sums up the pleasure of sorrel when it first springs forth in March 'in the making of sallets [it] imparts a grateful quickness to the rest as supplying the want of oranges and lemons. Together with salt, it gives both the name and the relish to sallets from the sapidity; which renders not plants and herbs only, but men themselves pleasant and agreeable.'

If the leaves are very small, no longer than 7.5 cm / 3 in, then throw them whole into the salad; if any larger, I would strip the leaf from the stalk. This has to be done if cooking the leaves, as the leaf effortlessly makes its own purée in a few minutes.

Melt butter in a pan and throw in the washed and dried leaf. Do not add any water or seasoning. Place over a low heat and within a minute, or at the most two, the leaves turn to a rather unpleasant hue of army-issue brownish-green. This is the only unattractive part of sorrel. However, the taste is stunning. For *Sorrel Sauce*, simply add to this purée some crème fraîche or fromage frais, which will lighten and improve the colour. For a theatrical improvement of colour, place some raw watercress leaves in a blender with the purée and liquidize. This sharpens the greenness and adds another dimension to the taste. It is a perfect sauce for fish of any kind, though for the more oily fish, like mackerel, an acid sauce like sorrel (without added watercress) is in my view a necessity.

For *Sorrel Soup*, reduce the sorrel leaves to a purée, then add raw watercress leaves and liquidize, before adding stock and single cream. This makes a wonderful cooling spring soup.

POTAGE À LA BONNE FEMME

This is my own version of the classic creamy soup.

30 g / 1 oz butter
450 g / 1 lb floury potatoes, peeled and diced
1 large onion, diced
handful of sorrel leaves (about 170 g / 6 oz)
1.75 litres / 3 pt vegetable stock
300 ml / ½ pt single cream
2 tbsp chopped chervil
2 – 3 tbsp crème fraîche
handful of croutons (optional)

Melt the butter in a saucepan and add the potatoes and onion. Let them sweat a little.

Tear the sorrel leaves from their stalks and add the leaves with the stock. Simmer for 20 minutes.

Leave to cool, then liquidize half of the soup, so that it is both smooth and textured.

Reheat and add the single cream. Serve with chopped chervil sprinkled on top and the crème fraîche and croutons, if using.

THE ORIGINAL JULIENNE SOUP

As wood sorrel has spread across the length and breadth of the British Isles, this soup could be made in early spring and summer. You will find the plant growing in woods, under hedgerows, in shady spots and on high ground. Pick a generous handful of the leaves complete with their stalks. Chop the stalks into 2.5 cm / 1 in lengths. Melt some butter and sweat some chopped onion, then add the stalks and leaves of the sorrel. Add a few pints of good vegetable stock and simmer for 10 minutes. Season to taste.

LAYERED FISH TERRINE

1.1 kg / 2½ lb mixed white fish high in gelatine,
such as brill, sole, angler fish, rock fish or flounder
1 carrot
1 onion
2 celery stalks
1 bay leaf
2 garlic cloves
juice and zest of 1 lemon
675 g / 1½ lb sorrel leaves
30 g / 1 oz butter, plus more for greasing
2 egg yolks
generous handful of chopped parsley
2 tbsp chopped chervil or tarragon
sea salt and freshly ground black pepper
sprigs of chervil, dill or fennel for garnish

Have the fishmonger fillet the fish, but ensure that you keep all the heads and bones. Place these in a saucepan with the carrot, onion, celery, bay leaf, garlic cloves and lemon zest. Bring to the boil and simmer for 20 minutes. Strain and leave this stock to cool, then add the lemon juice.

Meanwhile, preheat the oven to 200°C/400°F/gas6. Tear the sorrel leaves from their stalks and cook in the butter for a few minutes or until puréed. Set aside to cool. Mix in the egg yolks and season.

Chop the white fish into chunks and mix in the parsley and chervil or tarragon. Prepare a terrine dish by buttering the base and sides, then place a strip of buttered foil along the base and up both sides so that the finished terrine can be lifted easily.

Over the foil, place a sprig of chervil, dill or fennel, then pack a layer of white fish down to cover the base, pour in a little of the fish stock so that it just covers, then place a layer of the sorrel purée, followed by another layer of fish moistened with the stock and so on until the terrine is full.

Cover with buttered paper, place in a bain-marie half full of boiling water and cook in the oven for 30 minutes. Take out and leave to cool, then refrigerate for 4 hours before unmoulding.

Moisten a knife with boiling water and slide it around the sides, then gently pull the foil handles. Serve with a cucumber salad (see page 145).

Portulacaceae

PURSLANE

Portulaca oleracea

Members of this family often have fleshy leaves. Wild purslane (*Portulaca oleracea*) we treat as a common weed, though the best way of getting rid of it from the garden is to eat it.

You are likely to see purslane in bunches in large markets in the spring and early summer, much of it imported from Cyprus. It is yet another plant which used to be widely eaten – commonly grown during the Middle Ages and often pickled – which has now fallen into neglect. This is quite unfair, for it is succulent and very slightly peppery and the fleshy leaves give immense textural delight in a salad and upon the palate.

Gerard says, 'raw purslane is much used in salads, with oil, salt and vinegar.' Evelyn also praises it in salads, 'for it is eminently moist and cooling, especially the golden.' The golden variety with yellow leaves is less hardy than the green. I have failed to grow it in my garden, but green purslane I do grow, though it is best to start it off under glass. In the seventeenth and eighteenth centuries it was grown in hot beds, used when young in salads and the older plants pickled.

Purslane has had its critics too. William Cobbett, writes in his *The English Gardener* (1833), 'a mischievous weed, eaten by French men and pigs when they can get nothing else. Both use it in salad, that is to say, raw.' Earlier, however, the poet Robert Herrick (1591–1674) had written a charming verse:

Lord, I confess too when I dine
The pulse is thine,
And all those other bits that be
There placed by thee,
The worts, the purselain, and the mess
Of Water Cress.

Cruciferae

THE CABBAGE FAMILY

CABBAGE *Brassica oleracea* Capitata Group
BRUSSELS SPROUT
Brassica oleracea Gemmifera Group
KALE or CURLY KALE
Brassica oleracea Acephala Group
CAULIFLOWER *Brassica oleracea* Botrytis Group
SPROUTING BROCCOLI
Brassica oleracea Italica Group
CALABRESE *Brassica oleracea* Italica Group
KOHLRABI *Brassica oleracea* Gongylodes Group
TURNIP *Brassica campestrio* Rapifera Group
SWEDE *Brassica napus* Napobrassica Group
CHINESE CABBAGE or PE-TSAI
Brassica rapa Pekinensis Group
CHINESE MUSTARD GREENS *Brassica juncea*
PAK-CHOI *Brassica rapa* Chinensis Group
MIZUNA GREENS *Brassica rapa* var. *nipposinica*
WATERCRESS *Nasturtium officinale*
LAND CRESS *Barbarea verna*
GARDEN CRESS *Lepidium sativum*
RADISH *Raphanus sativus*
LARGE WHITE RADISH or DAIKON or MOOLI
Raphanus sativus spp.
BLACK RADISH *Raphanus nigra*
HORSERADISH *Cochlearia armoracia*
JAPANESE HORSERADISH *Wasabia japonica*
ROCKET *Eruca sativa*
SWEET ROCKET *Hesperis matronalis*
YELLOW ROCKET or WINTER CRESS
Barbarea vulgaris
SEAKALE *Crambe maritima*

This is a large family, containing arguably the largest number of edible varieties and characterized by a fairly general ability to withstand a rather cold climate. It also contains a number of well-known ornamental flowering plants (the cross shape formed by the four-petalled flowers gives the family its name), like scented stock and the wallflower, as well as the plant from which the Ancient Britons made the woad with which they painted themselves, and according to Caesar gave them 'a more terrible aspect in battle'.

The principal genus within the family is the brassicas, which numbers among its extensive ranks all the cabbages and their close relatives. Other important plants include the mustards, radishes, rocket, watercress and seakale, all noted for their pungent flavours.

CABBAGE

Brassica oleracea Capitata Group

We do not seem to be great enthusiasts of the cabbage today, though we feel more warmly about other members of the family. Yet writers in antiquity spoke highly of cabbage and its many medicinal qualities.

It was firmly believed that cabbage was a protection against drunkenness. Cato in *On Agriculture* sums it up: 'If you wish to drink deep at a banquet and to enjoy your dinner, eat as much raw cabbage as you wish, seasoned with vinegar, before dinner.' Athenaeus tells us that the Egyptians always began their meals with it. Furthermore the seeds, if eaten with the meal, would keep one sober. Cabbages had the quality of banishing headaches and, if the leaves had not been efficacious the night before, they would surely remove a hangover the day after. Cato also comments that the medicinal value of cabbage surpasses all other vegetables and gives five pages to its many cures.

Pythagoras loves the cabbage and Pliny comments that it would be a long task to make a list of all the praises of cabbage. It makes one surmise that the Ancients did not over-cook the vegetable so that it began to emanate that rank smell of bad institutional cooking. We would be wrong, however, as there are constant references to 'boiled cabbage glistening in oil' and some to the unpleasant smell of cooking cabbage. It was also believed that if you grew cabbages in a vineyard it made the wine dark. Athenaeus, quoting from a fragment by Nicander, tells us that they used to dry cabbage leaves (they still do in China today). 'You wash and dry them in the north wind and they are welcome in winter even to the idle...for cooked in warm water they come to life again.'

All cabbages stem from the wild variety *Brassica erratica,* which still grows on the coasts of Britain, France, Spain and Italy. The wild cabbage is the father of all kales, Brussels sprouts, cauliflower, broccoli and probably kohlrabi too. Both Pliny and Cato talk of three varieties: the curly, its leaves resembling parsley; another with a large stalk and open leaves; and a third with very small stalks, a smooth and tender cabbage. Visitors to Egypt from Greece complained that the cabbage grown there was inferior and demanded that seed be brought from Rhodes. Even so, the seed produced a bitter plant in the second year. There is some dispute as to whether this vegetable, which was so sought after, could really be the cabbage with which we are familiar today. The cabbage was one with a thick stalk – similar to the wild cabbage – and not the round-headed type.

By Roman times we have another clue as to the nature of the plant, for two types are mentioned: one, called *cymae* or *cauliculi* and sold in bundles, is expensive; while another, which sounds more like a type of kale growing near the sea, is a poor man's dish. Athenaeus writes 'They will boil sleek cabbages and serve pea soup with them.' Apicius mentions the bundles, of course, and Pliny tells us that the vegetables are tender young shoots that spring from the main stem. This sounds more like the Chinese family of brassicas, where the 'cut-and-come-again' type of culture brings a steady supply of young fresh shoots to the table. The gastronomic sensitivity which was so pronounced in the Ancient World, the Mediterranean climate and the short cooking times likely for summer fare imply that the much-loved cabbage was nearer the Oriental brassica than the cabbage we know today.

The headed varieties with which we are familiar were very likely to have been cultivated in Northern Europe by the Celts, who were passionate and efficient farmers. By 200 BC there were no wild woods left in Europe or in Britain itself, every available piece of fertile land was tilled and great quantities of grain were exported to continental Europe. Scholars usually so keen to praise the sophistication of the Classical World have tended to ignore the great achievements of the Celts.

With the Roman occupation of Britain and Gaul, both types of cabbage were brought to Northern Europe, where the kale-type – very similar to the wild cabbage – was taken up as a staple food. Historians tend to be disparaging about the 'ubiquitous cabbage', yet its highly healthy content – a concentrated package of several known therapeutic compounds – must have made it a valuable food in the Dark and Middle Ages. Pliny was right, after all, for there is little he claimed that cabbage does not cure.

Cabbages were chopped and added to marrowbone stews with leeks and onions and whatever herbs or spices were available, but saffron if possible. The first English recipe is from The *Forme of Cury*, where the cabbages are quartered, cooked in broth with onions and leeks and flavoured with saffron and salt. Much clarified butter also went into their cooking and this probably helps to explain how cabbage has always been over-boiled and served with butter. Traditions take long to die, especially when the main reason for them has vanished. Peasant medieval cooking was one-pot slow cooking, simmering all day over the embers of a fire.

Sauerkraut is thought to have been discovered by the Celts of Middle Europe. Harold McGee thinks that as it stems from the principle of pickled cabbage, it was brought to Europe from China by the Tartars, for they loved to salt foods to preserve them throughout the winter. Sauerkraut was certainly in use in the Middle Ages. 'He who sows cabbages and fattens a pig will get through the winter,' is a proverb which crops up throughout Europe. Sauerkraut preserves the cabbage's vitamin C, so for many centuries it would have been the only source of this nutrient in the winter months until the widespread use of the potato, which was not accepted until the nineteenth century. Sauerkraut, being a fermented product, is also rich in the benign bacteria which help and protect the colon.

After Pliny, however, cabbage did not acquire a good press. Richard Burton (1577–1640), author of *The Anatomy of Melancholy*, was extremely sour. 'Amongst herbs to be eaten I find gourds, cucumbers, coleworts, melons, disallowed, but especially cabbage. It causeth trouble-some dreams and sends up black vapours to the brain...'

Ever since, cabbage has sunk into the slough of diners' gloom, associated with poverty, over-boiling and 'a nasty history of being good for you,' as Jane Grigson puts it. Well, here are some details of its recent beneficial history.

NUTRITION

In 1931 it was found that rabbits survived a lethal dose of radiation if they ate cabbage leaves prior to exposure. Further tests in France in 1950 and in the USA in 1959 had similar results, in experiments with guinea pigs.

Furthermore, population surveys in Greece, Japan and the USA have linked cabbage with protection against colon cancer. 'One year-long survey in five areas of Japan published in 1986 concluded that those who ate the most cabbage had the lowest death rates from all causes...' It goes on to explain why, quoting work done by Dr Lee Wattenberg in the 1970s. Wattenberg isolated chemicals from the cabbage family, called indoles, that blocked cancer formation in animals. Cabbage and its whole family of brassicas guard cells against the first onslaughts of cancer.

However, an unbalanced diet based on great amounts of daily cabbage would not be wise, for cabbage in excess can aggravate thyroid problems and possibly cause goitres. Thiocyanate compounds peculiar to the cabbage family interfere with the body's uptake of iodine and possibly cause goitre, but this is only likely to happen in areas where the drinking water is also low in iodine.

Cabbage is an excellent source of vitamins A, C, B1, B2, B3 and D. It is high in iron, potassium and calcium. When cooked, however, it tends to lose about half its nutrients. Pickling does not destroy its vitamin C content; in fact, Captain Cook credited the good health of his crews to daily sauerkraut.

VARIETIES

Apart from the red cabbage, which is dealt with later, there are two main types of hearted cabbage – the Savoy and the Dutch white. The Savoy cabbage is the one much loved by painters, with its crimped and colourful leaves, for its striking aesthetic quality rather like a prize bloom from the Chelsea Flower Show. It is also, in my view, quite the tastiest of the cabbage varieties, having a delicate flavour which – if it is cooked briefly – is never ever sulphurous. It is a hardy winter plant and frost-resistant, so it is an important source of vitamins B and C throughout the winter season. Dutch whites are those pale-green smooth and glossy globes that have their leaves packed closely together. They keep particularly well.

Spring cabbage is the term used for young cabbage. The leaves are still open and the central bud has only just begun to form. It is sweet and very delicious in flavour.

COOKING

Cabbage is excellent raw, but if it is to be cooked let it be for the briefest time possible. For spring cabbage, simply strip off the coarse outer leaves and either cook the centre in a little salted water for 3 minutes, or slice and stir-fry in oil or butter.

Both the smooth white Dutch cabbage and the curly Savoy can be treated in a similar fashion. Especially in the winter, both can be grated for salads and tossed (or, better still, marinated for about an hour) in a strong mustardy vinaigrette – with or without the addition of garlic, chopped spring onions, sliced apple, celery or nuts. This kind of salad is very near to the commercial and much-despised coleslaw. Not that it deserves such denigration. It is, after all, a classic American salad, in which the cabbage is mixed with apple, carrot and a pinch of paprika. Bought coleslaw seems to have the addition of sour cream. The name stems from the Dutch *kool* (cabbage) and *sla* (salad), though others suspect that it stems from the medieval word for cabbage (coleworts).

As the Savoy cabbage looks so attractive the leaves can be blanched and used to wrap food in the manner of *dolmades* (see page 64). The Savoy is the classic cabbage for stuffing and cooking whole. I am not a fan of this dish, as it involves up to an hour's cooking and that old rank smell of boiled cabbage. The smell is caused by volatile sulphur-containing compounds, released when cooking begins to break down the cell structure. Once you begin to smell the brassicas, they have begun to cook too long. The amount of hydrogen sulphide produced from boiling cabbage doubles between the fifth and the seventh minute.

It is much better to stick to brief cooking times and the preservation of the nutrients. Because of its history in stews and winter soups there are hundreds of recipes which use cabbage in this way, but the cooking time in nearly all of them can generally be cut down to a few minutes – the leaves just need to be softened but not limp. One of my favourite ways with Savoy cabbage is to slice it across thinly and then stir-fry it with ginger, garlic and chilli.

Steaming cabbage is one method of ensuring that most of the nutrients and the flavour stay in the leaves. Quarter a Savoy, or if it is very large cut it into eighths, and steam for about 6-8 minutes. If in a hurry, pour 6 mm / ¼ in of salted water into a pan, bring to the boil, drop in the quartered cabbage and leave for 3 minutes. Then drain and serve.

CABBAGE AND GINGER SOUP

for 6

2 tbsp olive oil
1 tsp coriander seeds, crushed
50 g / 2 oz root ginger, peeled and grated
1 whole white Dutch cabbage, grated
1.75 litres / 3 pt vegetable stock
sea salt and freshly ground black pepper
2 – 3 tbsp sour cream, fromage frais or yogurt
few fresh chives or coriander leaves

Heat the olive oil in a large pan and throw in the coriander and ginger. Sweat the spices for a moment.

Add the grated cabbage and stir so that is coated and begins to shrink a little. Then add the stock, bring to the boil and simmer for 5 minutes.

Leave to cool, then blend to a smooth light purée. Taste and season.

This soup can be eaten cold or hot. Serve with some sour cream, fromage frais or yogurt spooned on top and sprinkled with a few chopped fresh green chives or coriander leaves.

SAUERKRAUT

Slice several whole white Dutch cabbages fairly thinly in a food processor, with a mandoline or (taking rather longer) with a knife.

Pack layers of it into a plastic bucket and season each layer with a few juniper berries, caraway or coriander and sea salt. Pack it tightly and place a weighted lid over the top. It is best then to leave it in a garage or outhouse as it tends to pong a bit.

Leave for 3 weeks while fermentation takes place. The sugars in the cabbage turn to lactic acid. The surface will become covered with a scum of yeasts and fungi, and these should be skimmed off. The liquid it has made can be topped up with salt water or wine.

Transfer the sauerkraut to sterilized jars, cover with juice and eliminate any air in the top of the jar. Screw down tightly and it will keep for years.

Before use, wash the sauerkraut thoroughly under cold running water to get rid of excess salt.

STIR-FRIED CABBAGE WITH WALNUTS

1 tbsp sesame oil
3 garlic cloves, chopped
1 red chilli pepper, chopped
1 tsp mustard seeds
1 whole Savoy cabbage, thinly sliced
1 onion, thinly sliced
8 g / ¼ oz chopped walnuts
3 tbsp dry sherry
1½ tbsp soy sauce
pinch of caster sugar
pinch of sea salt

Heat the sesame oil in a wok and throw in the chopped garlic, chilli and mustard seeds. Stir-fry for a moment until the mustard seeds begin to pop.

Then throw in the cabbage, onion and walnuts. Stir-fry vigorously for a minute or two, until the cabbage has shrunk by a third.

Throw in the rest of the ingredients, stir well and serve the dish immediately piping hot.

RED CABBAGE

All red, purple and blue plants – which include berries, currants, grapes, radishes, beetroots, aubergines and the red cabbage – acquire their pigmentation from anthocyanins. These are soluble in water, which is why all of these fruits and vegetables change colour when cooked and leak their pigmentation wherever they be. Once water touches it the red colour changes to purple or blue and the cabbage is, in fact, called *Blaukraut* in German.

Red cabbage, like all other brassicas, was developed by simply encouraging one element (in this case the pigmentation anthocyanin) over others in the original plant. It was grown in the Middle Ages. For years it was used in Britain for pickling in malt vinegar, a sad waste of a beautiful vegetable. Grated, it makes a particularly good salad with the addition of a strong garlicky vinaigrette. Other raw vegetables can be added, carrot for example, but you must be prepared for the colour to run. Red cabbage is rich in vitamin C. As a general rule, the darker the colour of the leaves the more vitamins it will contain. Thus outside leaves of plants will have more nutrients than the centre. Red cabbage is an exception, as it is dark in colour all the way through.

It is also an exception, I fear, in my rule above about cooking cabbages briefly. Raw red cabbage is excellent, but briefly cooked red cabbage does not have the flavour achieved by long slow cooking.

As a general rule, this flavour is brought out best by a sweet-sour addition, that is sugar or honey with wine, vinegar or lemon juice. The cabbage is braised for two to three hours over a gentle heat. Cooked like this it has long been a winter accompaniment to game; indeed roast pheasant and red cabbage is a classic combination.

SWEET-AND-SOUR RED CABBAGE

for 6

1 tbsp olive oil
5 garlic cloves, sliced
1 dried red chilli, broken up (optional)
1 onion, sliced
1 red cabbage, thinly sliced
5 tbsp red wine vinegar
2 tbsp muscovado sugar
1 tsp sea salt
freshly ground black pepper

Heat the oil in a large casserole dish and fry the garlic, chilli and onion for a moment.

Add the sliced cabbage and stir to coat it with the oil. Add the wine vinegar, sugar and seasoning. Stir again, place a tight-fitting lid on the casserole and let it cook over a very low heat for 2–3 hours.

I must say I prefer the cabbage when all its liquid is absorbed and it becomes a little sticky, so that some of it has caramelized. This is a matter of personal choice, but only this sort of long slow cooking will produce such a result.

BRUSSELS SPROUT

Brassica oleracea Gemmifera Group

Though the Romans valued cabbage sprouts as a great delicacy, this is now thought to be the new shoots sprouting from the central stem on the leaf axis, which was sold rather like asparagus in bundles. The sprouts with which we are so familiar were created by selectively breeding the stem cabbage for the tendency that it has to make budding heads at the junction of stem and leaves.

Its history, sadly, is rather obscure. Sprouts first appear growing around the area of Brussels in the Middle Ages, long before Brussels or Belgium actually existed. Then they seem to vanish entirely until a writer – Balehin in 1623 – talks of a plant bearing some fifty heads the size of an egg. Later, in 1699, we find a cooking instruction. 'Sprouts are very delicate, so boiled as to retain their verdure and green colour...' The gardening manual quoted by Dorothy Hartley goes on to add: 'The best seed of this plant comes from Denmark and Russia.'

So perhaps in those intervening centuries there are Russian or Scandinavian observations in a cookery or gardening book which could tell us more about this mysterious vegetable. For there is something odd about the plant, as if the cabbage has been tightly permed or over-wound up, so that its leaves are closely coiled, then shrunk and cloned. The Germans alone have brought a touch of poetry to the Brussels sprout; they call them *rosenkohl* – rose cabbage.

CHOOSING

Do buy Brussels sprouts with care, avoiding any with yellowing exterior leaves for they almost certainly have been lying around for a week or more. Buy only tight small buds and never any loose-leaved ones. Around Christmas-time you can often buy them still attached to their stem, and these will stay marginally fresher than the loose ones.

PREPARING AND COOKING

Constance Spry, agreeing with the gardening manual mentioned above, sums up their cooking. 'They should be regarded, I think, as minute and delicate little cabbages, mostly heart.' Indeed, as for cabbages, they need very little cooking and I, for one, enjoy them raw.

Though sprouts are still classed as *Brassica oleracea* (meaning a leafy garden vegetable used in cooking), their flavour is quite different from the parent plant. They are worth using raw in a variety of ways. Serve them as a salad with a vinaigrette. Slice them thinly with sliced onion or leek, or simply quarter the small ones and enjoy them raw in a plate of crudités.

If you choose to eat them cooked, then they can also be grated, stir-fried for a minute in olive oil or butter, then quickly seasoned and served. The small green buds are, after all, appealing to look at in themselves, so why destroy their shape? So if you want your Brussels sprouts cooked in the traditional manner, then throw them in a small amount of boiling

ABOVE Sweet-and-sour red cabbage served with roast pheasant

salted water and let them cook for 3 minutes – no longer. They should still have plenty of crunch to them and have all their flavour. If the sprouts are too firm for some cooks, leave them in for another 2 minutes; but once you detect a whiff of sulphur, they are – in my opinion – beyond repair.

I am astonished at how cooks of the past, distinguished for their taste in so many things, treated the poor Brussels sprout. Mrs Leyel, for example, in *The Gentle Art of Cookery*, insists that sprouts should boil for no less than 15 minutes. There are far worse stories than that: an aunt of a young friend of mine still puts the sprouts on to simmer before she goes to church on Christmas morning.

Lady Sysonby in 1935 in a collection of recipes used by her grandmother and mother has this to say: 'Brussels sprouts – small ones – after they are boiled and well drained, tossed up in butter in the frying pan with little bits of bacon or fried onion, make a dish everybody wishes to have more of.'

Let us play with variations on this idea. Allow only 2 minutes for the boiling of 450 g / 1 lb of small sprouts. Drain them well, toss in a frying pan coated with walnut oil together with the addition of 2 teaspoonfuls of mustard seeds and 2 oz / 50 g of chopped walnuts; or, toss in sesame oil with chopped garlic and chilli.

Constance Spry gives a recipe for a Brussels sprout purée which uses two parts sprout to one of potato purée, while Dorothy Hartley mentions that the very small sprouts are delicious 'scattered in broth'. So why not add these either to a clear soup or as part of a miscellany of winter vegetables for a hearty minestrone type of soup, remembering to add them only in the last 2 minutes of simmering. Josceline Dimbleby made famous a purée of Brussels sprouts (without any potato), emphasizing its bright green colour by serving it with a carrot purée.

One of the flavourings most complementary to the brassicas is the orange, as well as that of its cousins, and citrus jams and jellies. The recipes below explore this happy conjunction.

STIR-FRIED SPROUTS WITH GINGER AND ORANGE

1 tbsp sesame oil
30 g / 1 oz root ginger, peeled and grated
juice and zest from 1 large orange (about 3 tbsp)
450 g / 1 lb Brussels sprouts, trimmed and halved
½ tsp Tabasco
½ tsp caster sugar
pinch of sea salt
1 tbsp Curaçao or other orange-flavoured liqueur
2 or 3 spring onions, chopped
little chopped coriander leaves, to garnish

Heat the oil in a wok or frying pan and throw in the ginger and orange zest, followed by the sprouts. Stir-fry over a high heat for 1 minute.

Add the orange juice and Tabasco, continuing to move the sprouts in the pan for another minute or until the juice has almost evaporated.

Add the sugar, salt and liqueur. Give a final stir, so that the sprouts are covered with the juices. Add the chopped spring onions and fry for another 5 seconds or so.

Serve sprinkle with the chopped coriander.

SPROUTS POACHED IN LIME JUICE

3 limes
good pinch of sea salt
1 tsp caster sugar
450 g / 1 lb Brussels sprouts, peeled and trimmed
1 tbsp malt whisky

Take the zest from the limes, squeeze the juice from them and place in a saucepan. Add the salt and sugar and bring to the boil.

Throw in the Brussels sprouts and bring to a simmer. Place a lid on the pan and leave to simmer gently for 2 minutes.

Remove the lid from the pan and raise the heat. Shaking the pan so that the sprouts do not stick, boil rapidly until the liquor has evaporated to only a tablespoon or a little more.

Pour over the malt whisky and serve.

SPROUT AND NUT LETTUCE ROLLS

These tasty little rolls are excellent as appetizers or as a simple first course.

450 g / 1 lb Brussels sprouts, grated
30 g / 1 oz pine kernels, roasted
5 spring onions, finely chopped
juice and zest from 1 lemon
1 tsp Tabasco
1 tbsp walnut oil
1 tsp sugar
pinch of celery salt
10 or 12 lettuce leaves, blanched

Place the grated sprouts in a mixing bowl. Add the pine kernels and chopped spring onions, the juice and zest from the lemon, the Tabasco, walnut oil, sugar and celery salt. Mix thoroughly and leave to marinate for an hour.

Drain carefully, then squeeze the mixture to remove any remaining moisture.

Smooth out the blanched lettuce leaves, lay a tablespoon of the filling in one corner and roll up, tucking in the sides.

KALE or CURLY KALE

Brassica oleracea Acephala Group

The kale family are hardy thick-stalked winter plants with leaves of various shapes and types coming from the central stem (the Latin variation *acephala* means headless). Most people know the curly kale with leaves quite finely crimped like parsley. There are also silver and purple variegated leaves, some of which are used for floral arrangements. The large plain-leaved kales are still primarily used to feed livestock, as many people find the flavour of the leaves too strong for their palates.

Kale should not be confused with sea kale (*Crambe maritima* – see page 129) an entirely different plant, the young blanched shoots of which are eaten. If you grow kale yourself you have the opportunity of picking the tender new shoots and these are well worth eating, just steamed or lightly boiled for a moment. It is a great shame that farmers tend to be so conventional, for kale is very easily grown in our climate, but it is largely grown to feed cattle. If the young shoots could be harvested in winter and early spring a new luxury vegetable could be marketed.

Great quantities of kale were eaten in the Middle Ages, when it was cooked with oats, barley and onions to make a gruel as the main meal of the day. The basic recipe still appears in Mrs Beeton as kale brose – a Scotch recipe, where the ingredients make enough for ten persons: half an ox head or cow heel, a teacupful of toasted oatmeal, salt to taste, two handfuls of greens, three quarts of water. One can imagine this hearty broth being left to simmer all day and then providing the main meal for an agricultural labourer's family in the evening. She says kale needs four hours of cooking and is seasonable all the year round, but especially suitable in the winter.

GROWING AND VARIETIES

If you are a gardener, it is not difficult to have a vigorous supply of kale in the winter and early spring. Seed catalogues list over twenty different varieties, from Dwarf Green Curled to Westland Winter Toga. A word of warning, though; the young plants are an especial delight for caterpillars and a whole planting can be well-nigh wiped out overnight.

CHOOSING, PREPARING AND COOKING

If you are not a gardener and cannot grow your own, your only chance of buying kale is probably at a farm shop in early spring. It is one of the strongest-tasting of all the brassicas and goes well with mashed potato and pulses, but it is also enormously delicious as a vegetable in its own right.

First tear the kale leaves from the stalks and discard the stalks, then boil the leaves in a little salted water for 3–5 minutes and drain them well. We eat kale cooked in this simple manner frequently in the winter to accompany game. The cooked leaves can also be mixed with equal amounts of mashed potato and a diced onion, then seasoned with lots of black pepper and formed into small cakes which are egged, breadcrumbed and shallow-fried. Traditionally these go well with bacon or ham, though they are excellent with cold game.

Kale can also be added to soups and winter stews in a rather more sophisticated version than Mrs Beeton's Scotch recipe above. Being of fairly strong flavour, it goes well with beans and root vegetables. Because of its strength, kale will stand up to strong and pungent spices, so it will happily feature in many Indian curries.

KALE DHAL

If you can't find kale, spring greens will do just as well in this recipe.

1 tbsp mustard oil
1 tbsp mustard seeds
2 dried red chilli peppers, crushed
5 garlic cloves, chopped
2 tsp ground cumin
2 tsp ground coriander
2 tsp turmeric
zest and juice of 1 lemon
450 g / 1 lb kale, chopped small
450 g / 1 lb potatoes, peeled and chopped
850 ml / 1½ pt vegetable stock
115 g / 4 oz brown lentils
1 tsp sea salt

Heat the oil in a large saucepan and add the mustard seeds, chillies and garlic. Fry until the seeds pop.

Add the cumin, coriander and turmeric, together with the zest and juice of the lemon. Quickly add the kale and potatoes and move the vegetables around in the spices, then pour over the stock, bring to the boil and let simmer for 15 minutes.

Add the lentils and cook for another 20 minutes. Add the salt just before serving.

KALE or CURLY KALE CABBAGE *Brassica oleracea* Acephala Group

KALE AND POTATO PIE

This dish might seem excessively simple, but such a recipe allows the clear uncluttered flavour of kale to come through.

30 g / 1 oz butter, plus more for greasing
675 g / 1½ lb kale
30 g / 1 oz plain flour
300 ml / ½ pt milk
450 g / 1 lb potatoes, boiled and mashed
sea salt and freshly ground black pepper

Preheat the oven to 230°C/450°F/gas8 or preheat a hot grill. Butter a heatproof dish.

Cook the kale in a pan of boiling salted water for 3–5 minutes, then drain well. Layer the bottom of the prepared oven dish with the drained kale.

Melt the butter in a pan over a low heat, add the flour and stir in. Cook gently for a few minutes to make a roux, then stir in the milk to make a smooth white sauce. Season and pour this over the kale and then smooth over the topping of mashed potato.

Place in the oven or under the grill just long enough for the top to brown.

CAULIFLOWER

Brassica oleracea Botrytis Group

It is perhaps difficult to imagine the cauliflower as part of the cabbage family, but it is still *B. oleracea*, with the addition of the tag *botrytis*. This means 'clusters' or 'grape-like', which is an apt enough description of the large white curd. The cauliflower is, in fact, a cabbage which was bred for its flowers – a process that must have taken some time, for it made a late entry into Europe.

Originally it is thought to have come from China via the Middle East. It was brought to Spain by the Moors in the twelfth century, but arrived much later in Italy – not until the fourteenth century. Certainly the seventeenth-century writer and gardener Castelvetro took pleasure in cauliflowers and recommends a simple salad of the briefly cooked florets dressed with olive oil and salt. He also recommends that they should be cooked in broth and served on slices of bread with more broth poured over, seasoned with grated mature cheese – the first mention of cheese with cauliflower. The idea took another two hundred years before it became a popular dish in eighteenth-century England.

In earlier recipes cauliflower tended to be cooked in milk, as in this recipe from John Murrell's *Delightful Daily Exercises for Ladies and Gentlemen* (1621): 'How to butter a collflowre. Take a ripe collflowre and cut off the buddes. Boyle them in milk with a little mace and then pour them into a cullender and let the milke run cleane from them, then take a ladle full of creame being boiled with a little whole mace, putting to it a ladle full of thick butter, mingle them together with a little sugar, dish up your flowers upon suppets, pour your butter and creame hot upon it, strowing on a little slicst nutmeg and salt and serve it to the table hot.'

The idea of cooking cauliflower in milk died out slowly. Hannah Glasse is still recommending it in 1747, but by Mrs Beeton's day it is not mentioned.

NUTRITION

Cauliflower is particularly rich in vitamins and minerals. Like the rest of its family, it is also thought to be associated with lowering cancer rates. It contains phosphorus, sulphur, calcium and sodium and vitamins A and C – an average helping of cauliflower would supply half the daily recommended vitamin C requirements. It also contains considerable amounts of folic acid, particularly vital for pregnant women, as a deficiency of folic acid is the source of spina bifida in babies.

PREPARING AND COOKING

There has always been some dispute as to whether to cook cauliflower whole or broken up into its florets. Victorian recipes state that the whole cauliflower should be boiled for 20 minutes. I am not really an enthusiast of serving cauliflowers whole, as they are then either over- or under-cooked, while their beauty is destroyed at once at the table when the vegetable is broken up in order to be served. I do make an exception, however, in the early spring when it is sometimes possible to buy very small baby cauliflowers just the right size for an individual serving. These could not look more appealing when cooked for a dinner party.

The method I use is to arrange them in a baking tray containing 1 cm / ½ in of salted boiling water and place this in a preheated hot oven for 5 – 8 minutes, when they will both bake and steam. Drain them well and serve sprinkled with a mixture of equal parts of ground almonds and brown breadcrumbs which have been fried in olive oil and garlic, then seasoned and mixed with chopped parsley.

Because of its pleasant blandness, cauliflower has classically been teamed with a cheese sauce, possibly spiced with a little mustard or, as in Welsh rarebit, made with ale. Italian dishes have also often fried the florets in oil, garlic and chillies, then the dish is sprinkled with breadcrumbs and it is briefly grilled. Another Italian recipe adds anchovies and capers and this makes an attractive dish for lunch accompanied by a green salad.

The creamy blandness of the cauliflower also makes it an excellent and versatile foil for many stronger-flavoured vegetables, as in the selection of recipes that follow.

CAULIFLOWER SOUP WITH RED PEPPER GINGER SAUCE

This soup may be served cold as below, or both soup and sauce may be reheated and served hot.

1.75 litres / 3 pt vegetable stock
pinch of grated nutmeg
pinch of ground coriander
1 cauliflower, coarsely chopped
300 ml / ½ pt single cream
sea salt and freshly ground black pepper
for the red pepper ginger sauce:
2 tbsp olive oil
3 red peppers, deseeded, trimmed and chopped
1 red chilli pepper, chopped
55 g / 2 oz root ginger, peeled and grated

Bring the stock to a simmer in a large pan. Add the nutmeg, coriander and seasoning.

Add the chopped cauliflower and simmer for 20 minutes. Leave to cool, blend to a smooth purée, then add the cream.

While the cauliflower is cooling, make the red pepper ginger sauce: heat the oil in a pan, throw in the red peppers, chilli, ginger root and some salt. Put the lid on the pan and leave to cook over a very low heat for 20 minutes.

Let cool and blend to a thin sauce which will emulsify a little.

To serve, pour the soup into individual soup bowls. Put the pepper sauce in a jug and pour it carefully into the soup, starting in the centre and making a thin spiral. Serve at once.

CAULIFLOWER AND MUSHROOM PIE

shortcrust pastry bottom to line a pie dish, pre-cooked,
plus pastry for lid (see Swiss Chard Pie, pages 52-3)
1 cauliflower, broken up into florets
30 g / 1 oz butter
2 tbsp oil
2 garlic cloves, sliced
115 g / 4 oz mushrooms, sliced
1 tbsp plain flour
55 g / 2 oz dried fungi (porcini), rinsed and soaked in just enough water to cover
300 ml / ½ pt vegetable stock
small wine glass (5 tbsp) of vermouth or dry sherry
small wine glass (5 tbsp) of light soy sauce
sea salt and freshly ground black pepper

Preheat the oven to 200°C/400°F/gas6.

Bring about 5 cm / 2 in of salted water to a boil in a pan. Throw in the cauliflower florets, bring back to the boil and cook for 2 minutes – no more. Drain immediately in a colander and reserve.

Melt the butter with the oil in another pan. Add the garlic and sliced fresh mushrooms, and fry for a few moments. Then add the flour and stir over a gentle heat to make a roux.

Add the fungi with their soaking water, together

with the vegetable stock. Bring just to a simmer and cook for a few moments, then add the vermouth or sherry, soy sauce and seasoning (with this quantity of soy sauce you may need little or no salt).

Pour some of this mushroom sauce over the pastry base. Arrange the cauliflower florets in the pastry case, packing them tightly in. Then pour the rest of the sauce into the pie case. Roll out the pastry for the lid and fit it over the top. Cut a small vent in the centre of the top to allow steam to escape.

Bake in the oven for 30–40 minutes.

SPICY CAULIFLOWER ROULADE

This makes a perfect first course or an excellent supper dish with a green salad.

1 cauliflower, chopped very small
handful of finely chopped parsley
150 ml / 5 fl oz fromage frais
225 g / 8 oz frozen puff pastry, defrosted
1 large onion, finely chopped
1 egg, beaten
2 tbsp sesame seeds, toasted
for the hot sauce:
4 tbsp olive oil
2 red or green chilli peppers, finely chopped
1 red pepper, cored, deseeded and sliced
4 garlic cloves, peeled and sliced
1 tsp ground coriander
1 tsp ground cumin
pinch of sea salt
juice from 1 lemon
3 tbsp breadcrumbs, toasted

Preheat the oven to 220°C/425°F/gas7.

Add the cauliflower to a little boiling salted water and boil vigorously for about 3 minutes, then drain and reserve. (The cauliflower should never be mushy, the taste is then disgusting; nor should it be only one remove from raw for this recipe. Remember that the cauliflower pieces must be small and just soft enough to be rolled in the pastry.)

When the cauliflower is cool, mix in the parsley and fromage frais and reserve.

To make the hot sauce: heat the olive oil in a pan and throw in the chillies, red pepper and garlic. Cook for 2–3 minutes, then add the coriander, cumin and salt. Leave to cool. Add the lemon juice, then blend to a purée. Pour into a bowl and add the breadcrumbs to make a thick chunky sauce.

Roll the pastry out into a square 30 × 30 cm / 12 × 12 in, then spread the hot sauce over it to about 1 cm / ½ in from the edge. Sprinkle the chopped onion evenly on top of the sauce and then spread the cauliflower over that.

Roll the pastry up carefully and press down the ends to seal. Brush the ends of the pastry with a little of the beaten egg to seal properly. Brush the roulade with more beaten egg to glaze and sprinkle with the sesame seeds.

Bake in the oven for 25 minutes, or until the roulade is puffed up and golden. Leave to settle for 2 minutes before slicing.

CAULIFLOWER AND QUINCE GRATIN

zest and juice of 2 oranges
2 quinces
1 cauliflower, cut into florets
pinch of sea salt
2 tbsp good marmalade

Pour the orange juice and the zest into a saucepan, add the sliced quinces and then the cauliflower with the salt. Bring to the boil and simmer for 5 minutes.

Take out the cauliflower, draining well. Arrange in a warm heatproof dish.

Preheat a hot grill. Stir the marmalade into the quince and what is left of the orange juice, bring to the boil and pour over the cauliflower.

Place under the hot grill for a moment just to brown slightly.

RIGHT *Cauliflower and mushroom pie (see page 97)*

SPROUTING BROCCOLI

Brassica oleracea Italica Group

This vegetable gets its name from the Italian plural diminutive of *brocco*, meaning a sprout or shoot. It is a near cousin to the cauliflower and, instead of producing a single flower head, produces a cluster of loose flowers on several branches at the top of the stem, while below a large number of smaller heads grow in the axils of the leaves. Both white and purple varieties are grown, although there is no difference in the flavour and the purple variety turns green when cooked (see page 91).

The Romans were exceedingly fond of sprouting broccoli. It was sold in bundles and was expensive, for Pliny comments how they are table luxuries. Apicius gives several methods of serving the vegetable; it is to be cooked or served with cumin and salt, or with pepper, lovage, mint, rue and coriander leaves. The sprouting broccoli leaves themselves can be chopped and cooked in with the sauce, using *liquamen* (see page 143), oil and wine. He suggests another method is to mince the leaves with coriander, onion, cumin, pepper and oil, serving them with boiled leeks, olives or pine kernels and raisins.

BELOW *Sprouting broccoli*

Broccoli takes a long time to grow – a full year. It is planted in March and gathered the following spring. Because it has to over-winter, the species needs to be hardy and perhaps the intensity and strength of flavour is part of this characteristic. The milder Mediterranean winter of southern Italy would have helped its survival as a garden vegetable through the Dark and Middle Ages.

Castelvetro mentions it as 'the tender shoots which grow on the stalks of cabbage and cauliflower plants left in the garden over the winter'. He advises they should be cooked and served cold with oil, salt and pepper, adding they are much improved by being cooked with a few cloves of garlic. Anna del Conte observes that all the Italian recipes come from the south, where the broccoli is blanched and then cooked in olive oil, garlic and chilli to which soft breadcrumbs and cheese are added. There is no doubt that this is a most delicious recipe. She goes on to tell us that a Calabrian recipe puts all these ingredients in layers in a hot oven for 30 minutes. This, for my taste, is too long; the flower heads will have turned mushy and the strong flavour will be slightly rank.

John Randolph, who mentioned broccoli in 1775 in his *A Treatise on Gardening by a Citizen of Virginia*, summed up the essence of this delicious vegetable when he said: 'The stems will eat like asparagus and the heads like cauliflower'. Earlier, in 1695, John Evelyn comments 'The broccoli from Naples...are very delicate, as are the Savoys, commended for being not so rank but agreeable to most palates.'

In *Adam's Luxury and Eve's Cookery* (1764) we get very precise instructions on the cultivation but little on the cooking of broccoli. The authors first of all recommend the Roman and the Blue broccoli as being the best; they then tell you to sow it in May in a bed of good moist earth and, after instructing the gardener throughout the year, the plants are ready about Christmas or a little after, when the heads like small cauliflowers are cut off. These, Adam tells us, are soon succeeded by a great number of side shoots, which should be cut when about six or eight inches long. Eve suggests cooking the broccoli with a few cloves and serving it in a butter sauce.

It is these side shoots in early spring which I consider to be the most delicious. Here is the simplest method of cooking them. Take off the leaves and peel the stalks, tie them in a bundle and place them upright in some boiling water so that, like asparagus, the heads steam and the stalks are boiled. Cook for only 4 – 5 minutes. It obviously helps greatly if you have an asparagus kettle for the job.

On the other hand, if you do not mind the stalks fairly *al dente* (and I do not), steam the whole vegetable for 5 minutes. Serve with a lemon butter sauce (melted butter with half a lemon squeezed into it) or cook the florets briefly while peeling and eating the stalks raw, dipped perhaps in some sesame salt. The whole of broccoli, of course, after being prepared, can be eaten raw in a mixed platter of crudités and very good it is too, dipped into an *aïoli* mayonnaise, rouille or a herb-flavoured yogurt, smetana or fromage frais.

Doing anything more complicated with these tender sprouting shoots seems uncalled for, but I would welcome a stir-fry.

STIR-FRIED BROCCOLI WITH CHILLI AND PICKLED GINGER

2 tbsp sesame oil
2 garlic cloves, thinly sliced
1 dried red chilli pepper, crushed
55 g / 2 oz root ginger, peeled and thinly sliced
450 g / 1 lb broccoli, separated and stalks peeled and trimmed
55 g / 2 oz pickled ginger
pinch of sea salt
pinch of sugar
small glass (5 tbsp) of vermouth or dry sherry
sprinkle of light soy sauce

Heat the oil in a wok and throw in the garlic, crushed chilli and sliced root ginger. Stir-fry for only a second or two. Then add the broccoli and the pickled ginger. Stir-fry for 2 – 3 minutes.

Add the salt, sugar, vermouth or sherry. Continue to stir-fry for half a minute, then add the soy sauce and serve.

BROCCOLI POACHED IN ORANGE JUICE

1 large head of broccoli in one piece, weighing about 450 g / 1 lb
zest and juice of 2 oranges
1 tsp Tabasco sauce
½ tsp sea salt
½ tsp sugar

Slice the broccoli from the stalk so that the flower head is all of one piece, peel the stalk and slice it lengthwise into pieces.

Put the orange zest and juice in a saucepan and add the Tabasco, salt and sugar. Add the slices of broccoli stalk and then carefully lay the head of broccoli on the stalks. Cover tightly, bring to the boil and simmer for 5 minutes. If the juice touches the flower head it will drain the green colour and bleach it beige, which is unsightly, so the flower head must steam. Remove the flower head from the pan.

Now, increase the heat and reduce by half the juice around the stalks. Pour these and the juices around the broccoli head and serve.

BROCCOLI WITH A SHALLOT, ORANGE AND YOGURT SAUCE

450 g / 1 lb broccoli
for the sauce:
30 g / 1 oz butter
5 or 6 shallots, peeled and chopped
zest and juice of 1 orange
pinch of sea salt
pinch of sugar
4 tbsp Greek yogurt or fromage frais

Prepare and cook the broccoli as described opposite.

Make the sauce: melt the butter in a pan and add the shallots and the orange zest. Cook over a gentle heat for 3 – 5 minutes. Add the orange juice and cook for another 2 minutes. Take from the heat, add the salt, sugar and yogurt and mix in thoroughly.

Have the broccoli arranged in a warm serving dish and spoon over the sauce. Serve at once.

CALABRESE

Brassica oleracea Italica Group

This is a vegetable which looks like a grander version of the sprouting broccoli. In a sense, it is merely the same vegetable. It is a case of commerce having 'tidied up' the vegetable, so that the form and colour are aesthetically highly satisfying – but the flavour is often less so. It seems to have been made for colour photography, to complement good china and slim models in glossy magazines. It is, alas, designer food made for display in the supermarket and not to satisfy the palate.

Calabrese is grown in great quantities to be frozen and packaged. It appears in Chinese restaurant cooking, often covered with oyster or black bean sauce from the bottle. It can be purchased covered in cling film and is often seen slightly yellowing and showing its age. In the last few years, however, growers have produced calabrese which has a great deal more flavour, but do make sure you buy it when it is at its freshest – when dark green and very firm. When fresh, both calabrese and broccoli have 5 per cent protein, so they are an excellent vegetable to include in a vegetarian or vegan diet.

BELOW *Calabrese*

Peel the stalk and follow the instructions for asparagus. Alternatively, dice the peeled stalk and break up the calabrese florets, then steam or boil for a few minutes.

I would be tempted to cook calabrese as in the method given below, which is also excellent for broccoli and cauliflower.

CALABRESE WITH ANCHOVY AND CAPERS

675 g / 1½ lb calabrese
2 tbsp olive oil
1 dried red chilli pepper, broken up
3 garlic cloves, chopped
1 small tin of anchovies in oil
2 tbsp capers
3 tbsp toasted breadcrumbs

Cut the florets of the calabrese into their individual pieces. Peel and dice the stalks. Throw all the pieces into a pan of boiling water, cook for 2 minutes, then drain carefully.

Heat the olive oil in a pan and throw in the chilli and garlic, together with the anchovies and their oil and fry for a moment.

Then throw the calabrese pieces in and fry for several moments in the oil and flavourings. Sprinkle the capers and the toasted breadcrumbs over the dish and cook for a few more seconds until the breadcrumbs have soaked up the oil and become crisp and slightly golden in colour.

ABOVE Green kohlrabi

KOHLRABI

Brassica oleracea Gongylodes Group

Kohlrabi is not a popular vegetable in this country, but it deserves to be. In the rest of Europe, Israel, China and India, it is a favourite vegetable, but we have resisted its charms. It is still in the *Brassica oleracea* category, though the kohlrabi bears the distinguishing addition of '*gongylodes*', meaning roundish or swollen. In Northern India it is called *ganth gobhi*, meaning 'knotted' or 'bulbous cauliflower'. The part that is eaten is the stem, which swells into a globe rather than the root.

Kohlrabi can grow to the size of a grapefruit, but will then tend to have a woolly interior, so it is far better if harvested at half that size. There are two varieties, green and purple, but both taste similar. They have a distinctive flavour, something of fresh water chestnuts with a little touch of radish and celeriac. They make a marvellous foil for other flavourings and, used in the manner below, I have fooled people into believing they were artichoke bottoms. Kohlrabi is rich in potassium, fibre and vitamin C.

The reason for kohlrabi's dismal reputation in Britain is partly due, I suspect, to the fact that our cookery writers have never enthused over it. Elizabeth David ignores it altogether, though *chou-rave* is prominent in French vegetable markets in the autumn. Mrs Beeton did not care for them, damning with faint praise: 'Not generally grown as a garden vegetable,' she says, 'but if used when young and tender it is wholesome...' Jane Grigson thought 'there are better vegetables than kohlrabi, and worse.' While Constance Spry allows it four lines of tepid enthusiasm in her best-known work.

ABOVE Kohlrabi savoury starter

So let me swim against the tide and enthuse on the vegetable's merits. Its flavour is quite different from, but just as good as, both parsnip and turnip. However, this flavour is utterly lost if it is overcooked. Firstly, it can be grated and eaten raw as a salad, mixed with a good garlicky vinaigrette or mayonnaise and treated rather like a celeriac salad. Spices can be fried in sesame oil and poured over the grated kohlrabi, then the salad is tossed thoroughly.

Kohlrabi is also particularly good as a salad or in a platter of crudités when par-boiled. Simply slice it and simmer the pieces for a couple of minutes, then leave in cold water for a while before draining and mixing with other raw or semi-raw vegetables.

Grated, it can be lightly cooked: stir-fried like grated courgette in a little flavoured oil or butter with some mustard seed thrown into the oil, perhaps, to give it a little spice. It can be sliced paper-thin with a mandoline and fried in hot oil like crisps, or it can be cut into chip-size pieces and fried until golden brown. It can be diced and fried with garlic, ginger, chilli and onion.

Yet its bulbous shape suggested to me that it could be also cut and used as a base or hollowed out and stuffed. Here are various recipes which show, I hope, how varied and adaptable this vegetable can be. The first can provide a base for a number of first courses.

Choose kohlrabi no bigger in diameter than 6 cm / 2½ in, peel them and slice into 1 cm / ½ in discs. Drop these into boiling salted water and poach for 3 minutes. Drain and dry the slices. Dice the smaller discs but keep the larger ones whole. Fry them in olive oil and garlic until just brown. These can be used as a base like an artichoke bottom for various additions piled upon them. The fried diced kohlrabi can be used as part of the topping mixture. I suggest mixtures of mushrooms, red and green peppers, sweetcorn, chillies, capers, olives, anchovies, in dressings of walnut or sesame oil.

Kohlrabi goes very well with the silky Greek sauce *avgolemono* – a mixture of lemon and chicken stock. For this it should be peeled and sliced, boiled briefly for 3–4 minutes, then drained carefully. It is placed in an ovenproof dish, the sauce poured over and the dish briefly placed beneath the grill.

If you want to use the vegetable whole, it is best stuffed. Peel the kohlrabi, then cut out a good hollow in the centre and boil the kohlrabi for 8 minutes. Drain with care and fill the centre with whatever stuffing you choose. Bake in a preheated hot oven for 20 minutes. All of the round root vegetables (turnips, swedes, sweet potatoes, yams, cassavas) can be given this treatment and they make good winter fare.

Kohlrabi can be used instead of potato in the great gratin dishes, Dauphinois, Lyonnais and Ardennais. In fact, the last, flavoured with juniper berries, works particularly well (see below). Kohlrabi can also be used instead of the cauliflower in the Cauliflower and Mushroom Pie (page 97).

KOHLRABI WINTER SALAD

2 kohlrabi, peeled and grated
3 carrots, peeled and grated
1 celery heart, chopped
bunch of parsley, chopped
55 g / 2 oz pine nuts
1 garlic clove, crushed
3 tbsp olive oil
1 tsp red wine vinegar
pinch of sugar
1 tsp Dijon mustard
sea salt and freshly ground black pepper

Thoroughly mix all the vegetables in a large bowl with the parsley. Fry the pine nuts with the garlic in the olive oil until golden. Add these to the salad.

Add the wine vinegar, sugar and seasoning to the oil in the pan, together with a good teaspoon of Dijon mustard. Stir well and pour over the salad.

Toss and leave to marinate for an hour.

KOHLRABI SAVOURY STARTER

1 tbsp olive oil
2 red peppers, cored and sliced
1 yellow or orange pepper, cored and sliced
1 red chilli pepper
5 garlic cloves, sliced
115 g / 4 oz mushrooms, sliced
55 g / 2 oz pickled ginger
bunch of rocket leaves
4 slices of kohlrabi to be used like artichoke bottoms (see left)
4 tbsp tapenade
sea salt and freshly ground black pepper
4 spring onions, finely sliced, or fresh coriander, chopped, to garnish

Heat the olive oil in a saucepan, throw in the peppers, chilli, garlic and mushrooms. Season, then cover and leave over a low heat for 20 minutes.

Add the ginger, stir and remove from the heat.

Take 4 plates and arrange some rocket leaves on each. Place a kohlrabi disc in the centre of each rocket bed and spread each of those with the tapenade. Pile some of the pepper and mushroom mixture on top of that.

Garnish with sliced spring onion or chopped coriander and serve cool or warm.

CHOU-RAVE À L'ARDENNAISE

55 g / 2 oz unsalted butter, plus more for greasing
3 kohlrabi, very thinly sliced
2 large onions, thinly sliced
1 tbsp juniper berries
300 ml / ½ pt vegetable stock
sea salt and freshly ground black pepper

Preheat the oven to 200°C/400°F/gas6.

Butter a shallow earthenware oven dish and arrange in it alternating layers of kohlrabi and onion, throwing in a few juniper berries on each layer and dotting with a little butter and seasoning as you go.

Pour over the stock and place in the oven for 40–45 minutes, or until the top is golden brown.

ABOVE *Turnips with tomato and garlic*

TURNIP

Brassica campestris Rapifera Group

I believe our disdain for this delicious vegetable and the inane way we cultivate it – leaving plants to grow too big and too old – stems from their use in the eighteenth century as new cattle fodder for the winter months. Defoe, during his tour of England in 1722, remarked on the extent of the turnip cultivation and how fat the cattle now looked. Turnips and oil seed rape cakes had been used thus for many years in Holland, though it was the Romans who first used it as winter fodder and they may well have introduced the practice into Gaul. Turnips had the added blessing of growing well on poor soil and ignoring cold weather, so immense tracts of previously unproductive land came under cultivation.

The turnip is thought to have been indigenous to the area between the Baltic Sea and the Caucasus, where they were later cultivated and from where they spread into Europe. Its reputation was poor, however, and it was always linked with the food of the poor. The Roman soldier, farmer and writer Columella, points out in the first century AD that they are an important food for country people as they are so filling. The leaves were certainly eaten with as much pleasure as the roots. There are early recipes for pickling the roots and even the gourmet Apicius gives one where the turnips are sliced and pickled with honey, vinegar and myrtle berries. Pickled turnip, in fact, is delicious (see the recipe opposite). The Romans also served roast duck with roasted turnips, a dish which is still traditional in France, though it only appears in England as an Elizabethan recipe for boiling a duck with turnips. They were boiled separately, then turned in sweet butter with cinnamon, ginger, pepper, mace and salt and served with the carved bird. The young leaves of turnips were used in salads through the sixteenth and seventeenth centuries. Other salads contained the sliced turnip which had been boiled then dressed with oil and vinegar.

For the last five centuries or so the turnip has endured a hard time in Britain. It has been largely over-cooked, pulverized, mashed, pulped and puréed. Why can't we grow those tiny round purplish turnips that the French so revere? The answer is that we choose still to cultivate the turnips principally for cattle fodder and pick for human consumption the same variety just a little earlier.

The French turnip (*navet*) is *B. napus esculenta*; *esculenta* means edible and it certainly is. Sturtevant, who generally restrains his lyricism, is moved to say 'it surpasses other turnips by the sweetness of its flavour and furnishes white, yellow and black varieties.' These smaller turnips certainly crossed the Channel, for they are mentioned in 1683 and in a garden dictionary as late as 1807, but they never caught on. British cuisine was dedicated to over-boiling root vegetables and refused to see them in any other light. Fortunately, greengrocers and supermarkets now stock the small French turnip and we can enjoy them.

NUTRITION

If you are lucky enough to find turnip tops, cook the young leaves. They are particularly high in vitamins A, B and C. Boil for a few minutes, drain well and serve with a little butter or olive oil. Turnips themselves are 90 per cent water, with traces of vitamins and minerals. They are very low in calories.

CHOOSING AND COOKING

Buy turnips with healthy and bright green tops; the roots should be firm with no discolouring. Turnips are excellent raw, grated and then dressed as for a winter salad. They do not need peeling. Or they can be sliced in fingers as part of a platter of crudités.

Even when cooked, turnips should have some crunch. Boil them for 3 minutes, then slice them and cover with a garlicky vinaigrette or melted butter.

TURNIPS WITH TOMATO AND GARLIC

1 small tin of chopped tomatoes
2 garlic cloves, chopped
1 dried red chilli pepper, crushed
1 small glass (5 tbsp) of red wine
sea salt and freshly ground black pepper
450 g / 1 lb turnips, trimmed and thickly sliced
2 tbsp chopped parsley

Pour the tomatoes with their liquid into a pan with the garlic, chilli, wine and seasoning. Bring to a simmer and cook for a moment.

Add the turnips, mix them into the sauce and let them simmer quietly for 5 minutes.

Turn them into a warmed serving dish and sprinkle with a little parsley.

QUICK PICKLED TURNIPS

You will find this method of pickling turnips all over North Africa and the Middle East, but this particular recipe comes from the Lebanon.

The turnip turns a delicate pink and is excellent as an appetizer with drinks.

675 g / 1½ lb baby turnips, trimmed
575 ml / 1 pt cider vinegar
2 tbsp sea salt
1 small beetroot, peeled

Slice the turnips into circles about 3 mm / ⅛ in thick.

Bring the vinegar to the boil and add the salt.

In a pickling jar, arrange some layers of sliced turnip and place the raw beetroot cut in half in the centre. Fill the jar up with turnip slices.

Pour the vinegar over. If there is not enough, fill up with water. Place a tight-fitting lid on, then turn the jar over several times so that the contents are well mixed.

Leave for 4-5 days before eating.

SWEDE

Brassica napus Napobrassica Group

A very modern vegetable, the swede is believed to have originated from a hybrid between the turnip (*B. rapa*) and kohlrabi (*B. oleracea*), though the latter species comprises the cabbage and all its near relatives. The root is larger than the turnip and either white, yellow or orange. Some varieties have a slight purpling at the neck, but once peeled the root is uniform in colour. Despite its name, the swede came from Bohemia in the seventeenth century and was used as winter food for both cattle and man. It was introduced into England from Holland in 1755, and called the 'turnip-rooted cabbage'. It quickly became popular as another easy crop to feed cattle.

There is some confusion over the names: in Scotland a dish of turnips will be swedes, in America they call it rutabaga, which is French for the large yellow variety – while the other varieties have the prettier name in France of *chou-navet*. We changed from calling the vegetable 'turnip-rooted cabbage' to 'swede' when Sweden began to export their crop to us in the 1780s. Gastronomy has avoided the swede like the plague. Is this justified, I wonder?

Like the turnip, it is considered to be worthy of cattle, but not of man. Yet the flavour is sweet, nutty and very pleasant. It has one drawback: mashed swede always seems too watery to be very enticing. It is, however, an immensely useful vegetable for winter soups. Peeled and diced, then boiled with other vegetables, like turnip, potato and carrot, and perhaps some bones from game birds, the resulting soup is a hearty and flavourful mixture, not to be despised on a cold winter's day.

Swedes need very little cooking before they disintegrate. I suggest methods of cooking which keep their water content and most of their nutrition, for swedes are rich in vitamin C, calcium and niacin.

COOKING

Peel and slice the swede into 3 mm / ⅛ in discs or lengths. Fry in olive oil with garlic over a high heat, stirring the pieces in the pan for about 2–3 minutes. They need to shrink a little and to brown at the edges, so that they just caramelize. This cooking method keeps the swede crisp on the outside but juicy within. Finish with a little sprinkling of salt, a pinch of sugar and a good twist or two of the pepper-mill. A little grated Parmesan over the top before serving would also do them no harm.

Alternatively, use the fried swede discs, which are as large as coffee cup saucers, as a base for mushrooms in tomato, avocado and rocket, or any mixture you like as a first course.

Winter vegetables are economical; they are also warming and make hearty wholesome dishes, but these are not to be despised. There is huge satisfaction to be gained from simple peasant dishes, such as the one that follows.

SWEDE AND KALE PIE

450 g / 1 lb kale, trimmed and chopped
30 g / 1 oz butter, plus more for greasing
30 g / 1 oz plain flour
300 ml / ½ pt milk
55 g / 2 oz Gruyère cheese, grated
55 g / 2 oz Parmesan cheese, grated
2 tbsp olive oil
3 or 4 garlic cloves, sliced
1 large swede, peeled and sliced
sea salt and freshly ground black pepper

Preheat the oven to 200°C/400°F/gas6 and butter an oven dish.

Throw the kale into some boiling salted water and let it cook for 4–5 minutes. Drain thoroughly, pressing the water out to make sure all the excess has drained away.

Make the sauce by melting the butter in a pan over a gentle heat and stirring in the flour to make a roux. Add the milk and the cheeses until it makes a smooth thick sauce. Season and combine the sauce and the kale. Pour this into the oven dish.

Heat the olive oil and the garlic in a frying pan and fry the sliced swede for a few minutes so that it turns golden. Arrange the slices over the kale.

Bake in the oven for 20 minutes.

CHINESE CABBAGE or PE-TSAI

Brassica rapa Pekinensis Group

Also known as Chinese leaves, this variety of cabbage has now become quite a familiar sight. After bean-sprouts it is probably the most well-known of the Oriental vegetables in the West. It has its fans; for example, Joy Larkcom talks of its sweetness and delicacy and Ken Hom commends us to use it raw or lightly cooked, for it is 'a sweet, crunchy vegetable with a mild but distinctive taste that chefs like to match with foods with richer flavours.'

I fear that, unlike the other Oriental brassicas, I cannot view it with great excitement unless it is cooked in a specific Oriental manner. Hom is right, it is the richer flavours that it needs. To my taste, the leaves have one excellent quality: they are a good vegetable for pickling. The process of adding spices, salt and vinegar makes them palatable, as do other Chinese recipes.

Both in Szechwan and in Tientsin, they cut the leaves and stalks into small chunks. In the latter they pack the cabbage into jars with salt, garlic and other

BELOW *The harvesting of pe-tsai*

spices. In Szechwan they preserve the cabbage without garlic. Various recipes call for the use of pickled vegetables, which give a different texture and a sharp flavour to all manner of soups and stir-fried dishes.

The history of Chinese cabbage is as long as that of China itself. Cabbage is mentioned in the fifth century AD. It took over slowly from the mallow, which had been the most important leafy vegetable of ancient China. Now, alas, the venerable mallow has come to be regarded as a weed. Much later, a Nanking writer in an essay called 'Precious Things' written about 1600, describes such delicacies as lotus root, ginkgo nuts, shad and small carp, but does not omit the daily fare of vegetables and fruits, 'the turnips, onions, celery, cabbage and pears.'

John Barrow, the traveller and civil servant, said of China in the Chia-ching reign (1796–1820) that in the 'assortment of foods there was a wider disparity in China between the rich and the poor than in any other country of the world.' Sadly, cabbage again was indissolubly linked with the diet of the poor – or the ascetic, as in this sad tale from *The Scholars* by Wu Ching-tzu, written at the same time:

'As they were chatting, lights were brought in, and the servants spread the desk with wine, rice, chicken, fish, duck and pork. Wang Hui fell to, without inviting Chou Chin to join him and when Wang Hui had finished, the monk sent up the teacher's rice with one dish of cabbage and a jug of hot water. When Chou Chin had eaten they both went to bed. The next day, the weather cleared. Wang Hui got up, washed and dressed, bade Chou Chin a casual goodbye and went away in his boat, leaving the schoolroom floor so littered with chicken, duck and fish bones and melon-seed shells, that it took Chou Chin a whole morning to clear them all away, and the sweeping made him dizzy.'

The prevalence of cabbage in China for most of this century tells us that it is still a very important staple; with mustard leaves, radishes, rice and soy beans it is the diet of the peasant. The main meal is a bowl of rice, a mass of bean curd and a dish of cabbages, fresh in season or pickled. Cabbages yield enormous harvests and are available in southern China all year round (*pekinensis* can grow through the winter).

The Chinese cabbage family is vast. Over a dozen varieties are grown in California for the US market alone and more are grown in Israel for the European market, while many more varieties are grown in China itself. This profusion of Chinese cabbage types can make even botanists confused. L. H. Bailey, who has written two volumes on the brassicas, has admitted to being quite perplexed.

Chinese cabbage is high in vitamin A, B1, B2, B3, C and D, as well as having plenty of minerals, including iron, potassium and calcium. Through cooking it will lose about half of its vitamin content, but pickling does not destroy its vitamin C. There have been various studies around the world which link regular cabbage consumption with some degree of protection against colon cancer (see page 89).

CHOOSING AND COOKING

It is a simple matter to select the least damaged Chinese cabbage, for if they have been around too long the outside leaves get tattered and torn. There should be very little wastage in Chinese cabbage: the outside leaves get thrown away and then everything else can be used. Take off what you need at the time and keep the rest in the salad drawer of the refrigerator. Either slice across the leaves and stalk or cut down the leaf and slice into strips.

Stir-fry the leaves with powerful flavourings, like sesame oil, garlic, ginger and chilli. Finish by adding a little salt, sugar and a tablespoon of Chinese rice wine or dry sherry. If you wish, you can also add at the end a little oyster sauce or black bean sauce. The cabbage needs transforming, so do not be nervous about adding strong ingredients. Some other suggestions follow.

CHINESE CABBAGE WITH GINGER AND COCONUT

1 tbsp sesame oil
3 garlic cloves, sliced
1 tsp coriander seeds, crushed
55 g / 2 oz root ginger, peeled and grated
450 g / 1 lb cabbage leaves, chopped
150 ml / 5 fl oz coconut milk

zest and juice of 2 limes
2 tbsp pickled ginger slices
sea salt and freshly ground black pepper

Heat the sesame oil and fry the sliced garlic, crushed coriander seeds and grated ginger for a second or two. Throw in the chopped cabbage leaves and season. Stir-fry for 2 minutes.

Add the coconut milk, lime juice and zest, together with the pickled ginger. Place a lid over the pan and let it simmer for 2 minutes.

Pour into a dish and serve.

CHINESE CABBAGE WITH MUSHROOMS

2 tbsp groundnut oil
1 dried red chilli pepper, crushed or broken up
450 g / 1 lb mushrooms (oyster would be best)
1 large green pepper, cored and sliced
450 g / 1 lb cabbage leaves, chopped
1 bunch of spring onions, chopped
pinch of sea salt
pinch of sugar
2 tbsp dark soy sauce

Heat the oil in a wok and throw in the chilli, mushrooms, green pepper and cabbage. Stir-fry for several minutes, until the cabbage begins to soften.

Add the spring onions, salt, sugar and soy. Stir-fry for another minute and serve immediately.

PICKLED CHINESE CABBAGE

Piled up on a small dish, these pickles are excellent appetizers to eat with drinks – the perfect foil for dry martinis. They can also be added to stir-fried dishes.

1.35 kg / 3 lb Chinese leaves, chopped
450 g / 1 lb sea salt
3 heads of garlic, cloves peeled and sliced
2 or 3 red chilli peppers, chopped

Pack the cabbage tightly into a large earthenware crock or a strong glass jar, salting and adding garlic and bits of chilli to each layer. Pack down tightly and place a board and a weight over the top.

In 2 – 3 days a liquid will begin to ooze around the lid and the pickling has begun. It is ready to be eaten after 3 weeks. It will keep for 2 – 3 months or more, as long as the cabbage is submerged.

Take out what you need and rinse thoroughly under a tap to wash away excess salt, then squeeze the pieces of cabbage further to get rid of moisture.

CHINESE MUSTARD GREENS

Brassica juncea

This is a favourite vegetable of mine. I grow it every year and when I discovered this passage from Pliny I thought I had possibly found another enthusiast:

'...with its pungent taste and fiery effect [mustard] is extremely beneficial for the health. It grows entirely wild, though it is improved by being transplanted: but on the other hand when it has once been sown it is scarcely possible to get the place free of it, as the seed when it falls germinates at once. It is also used to make a relish, by being boiled down in saucepans till its sharp flavour ceases to be noticeable; also its leaves are boiled, like those of all other vegetables. There are three kinds of mustard plant, one of a slender shape, another with leaves like those of turnip, and the third with those of rocket. The best seed comes from Egypt.'

There is, of course, no way of telling whether the vegetable that is 'boiled down' is *B. juncea* or *B. nigra,* which gives us black mustard seeds. There are many varieties of mustard, some annuals, some biennials, all of which are thought to have originated from the central Asian Himalayas. Selective breeding of various types over the centuries has led to an astonishing diversity. (See Joy Larkcom's fine *Oriental Vegetables* for a summary of many of them.) Many members of the mustard family grew wild in the Mediterranean, but as *nigra* tends to grow to at least 2 metres / 6 feet high, Pliny would surely have mentioned it.

Incidentally, it is the seed of *nigra* which is used

for Dijon mustard. White mustard (*Sinapsis alba*) is the plant seedling grown and eaten with cress seedlings; it is milder in flavour to black and contributes to the taste of American mustard, while a little of the white seed is used in English mustard. *Juncea* has a seed which is a lighter brown and which is also used in the manufacture of mustards worldwide, as it can be harvested mechanically.

In India, mustard greens are also grown for the extraction of mustard oil, an important ingredient in Indian cooking. The oil was also an important export, even under the British Raj. In the year 1871–2 they exported 1,418 tonnes of mustard seed for the extraction of its oil. It was then used in Russia as a substitute for olive oil.

The delight of being able to grow mustard greens – they germinate quickly and easily in our climate – is that the seedling and young leaves can be picked for green salads. Mixed in with much blander leaves, such as lettuce, they give huge zest and interest to the salad. The larger leaves can be stir-fried in groundnut, sesame or mustard oil and then simply dressed with a little light soy sauce, sea salt and some lemon juice. The flavour is so delicious that they need very little else.

Do not be anxious, for once cooked they lose at least half of their fiery flavour. This is not a bad thing for the older mustard leaves are more pungent than the young leaves picked for salad.

PAK-CHOI

Brassica rapa Chinensis Group

To confuse matters, pak-choi (also spelt bok-choy) is also called 'Chinese celery cabbage' and even 'mustard cabbage'. It has thick white stems which fuse on the bottom half of the plant; the leaf is spoon-shaped and a glossy dark green. They look more like Swiss chard than cabbage, and their flavour is very slightly peppery – not in the least bland like the Chinese leaf, nor fiery like the mustard greens. They are, in fact, a lovely vegetable to add to the repertoire. They are easy to grow and are now fairly easily obtainable from supermarkets and Chinese stores.

LEFT Pak-choi stir-fried with garlic, chilli and ginger

They are as old as Chinese leaves in the cuisine of ancient China and there is a bewildering variety of them. Bruce Cost tells us that Hong Kong farmers alone grow over twenty kinds of bok-choy, with names such as 'horse's ear' and 'horse's tail'.

Like mustard greens, pak-choi is easily grown in Britain's climate. The leaves can be eaten as a salad when very small; either young leaves or thinnings, where the whole plant can be used. The stalks are juicy and sweet. Once the plant matures, use the leaves for stir-frying. If the plant begins to flower, all the better, as the flowers are delicious in salads or can be stir-fried with the leaves.

Bruce Cost recommends the Shanghai bok-choy to be cooked whole or cut in half lengthwise, as 'the brilliant green of the cooked hearts makes an attractive adornment to a cooked meat dish.' Pak-choi is at its best when stir-fried fairly simply, adding sesame or groundnut oil with garlic, chilli and ginger, finishing with a pinch each of sea salt and sugar.

MIZUNA GREENS

Brassica rapa var. *nipposinica*

You will find these delightful fronds in Japanese cooking rather than Chinese. Although they originally came from China, it is Japan that has cultivated them since antiquity. The aesthetics of the Japanese cuisine are more precise and detailed than those of the Chinese and I suspect this vegetable found its place in national taste because of its intricate beauty.

It is again very easy to grow in colder climes. It forms a bushy rosette of feathery dark green leaves and the leaf stalks are white but quite slender and juicy. The flavour is of mild pepper and nuts. I grow mizuna every year and would never be without it. Plants from a spring sowing will last through to around Christmas and still withstand cold and frost. The young leaves are excellent in salads and the older leaves are very good stir-fried.

WATERCRESS

Nasturtium officinale

There was a Greek proverb, 'eat cress and learn more wit', which summed up a belief in antiquity that it was brain food. The Romans went so far as to recommend its consumption with vinegar for the insane. But which cress were they extolling? It is a term which covers all the crucifers which have slightly pungent and peppery leaves. Cress was widely known in ancient times. 'Basil, sorrel, spinach, cress, rocket, orache, coriander and dill are plants of which there is only one kind, as they are the same in every locality and no better in one place than another,' announces Pliny. Some disagreed: Eubolus according to Athenaeus, declares that Milesian cress is the best. It is likely that the mixture of mustard and cress was certainly used in Greek times, and thus we know that here they must have been referring to land cress (see below). When the Greeks talk of cress used as seasoning or as an ingredient in a sauce, however, it seems more likely that they are referring to the cress that grows at the edge of streams. We find them praising cress used as a seasoning with 'young greens, capers, eggs and smoked fish'. And again 'the wild plants fit to cook are lettuce, cress, coriander and mustard'. It surely is watercress when Xenophon comments how 'pleasant it is to eat barley cake and some cress when one is hungry by a stream.'

BELOW *Watercress*

It was noticed early on that, though watercress had a peppery quality, it refreshed the palate and re-invigorated the spirits. There is a charming story of the French king, St Louis (Louis IX) out hunting on a hot day and calling for water. None was available but a servant brought him watercress, which so refreshed the king that he decided to honour the plant and the place where he had eaten it. The arms of the city of Vernon still today show the royal symbol, the three *fleurs-de-lis*, on one side and three bunches of watercress upon the other.

For a banquet to honour the Comte de la Marche, Charles VI's chef Taillevent wrote on the menu after the fourth course, 'the watercress, served alone, to refresh the mouth.' How dismal it is now to see watercress relegated to a garnish with steak, and dismissed to the edge of the plate.

The first attempt on any scale to cultivate watercress was in the sixteenth century at Erfurt in Germany, a principal centre of wood growing, for the extraction of its blue dye. Erfurt lies between the Harz mountains of the north and the Thuringian forest. Watercress needs clear fast running water to grow successfully. It was not until 1808 that a William Bradbury created the first watercress beds at Springhead in Kent. Freshly cut watercress was sold at markets as a breakfast food, with bread, for industrial workers. A Victorian, E. S. Dallas, comments, 'they are the one vegetable eaten by rich and poor in England, for health even more than for food.'

According to Dorothy Hartley, scientific advances in the eighteenth century made people realize the value of the 'iron-rich' greenery. She goes on to say: 'Very young cress is light bright green, but the knowledgeable prefer their cress "sunburnt" – that is slightly older – when the leaves have acquired a delicate bronze.'

The main season for watercress is from October to May, with a break in mid-winter if it is very cold. Watercress cultivation is based around freshwater springs at the base of chalk uplands. Watercress had always been valued for its colour; soups and sauces made from it look good and taste wonderful. We should experiment far more with it in this way (see below). When using it for cooking do not cut all the stems off: allow a little to stay, as the most pungent flavour is there.

NUTRITION

Watercress is rich in vitamins A, B2, C, D and E and contains many trace minerals, including manganese, iron, phosphorus, iodine, calcium, potassium, sodium and sulphur. It also has antibiotic properties similar to those of the onion family, so is a highly useful food to have as a staple in the diet.

WATERCRESS SOUP

Most recipes suggest cooking the cress in stock and adding cream, but cooking cress changes the colour and dampens the flavour. With a food blender, this soup is simplicity itself. How rich or thick you want the soup is up to you. You can add cream or yogurt to the stock and watercress mixture for a richer texture if you wish.

1 bunch of watercress, stems removed
575 ml / 1 pt vegetable stock
575 ml / 1 pt buttermilk, smetana, skimmed milk or soya milk
sea salt and freshly ground black pepper
few spoonfuls of fromage frais or Greek yogurt, to garnish

Reserving a few whole sprigs for garnish, place the watercress in a liquidizer with the stock and blend until you have a smooth light green purée.

Mix whatever type of milk you choose with it and then season to taste.

If you want to serve the soup cold, leave it in the refrigerator for an hour or two to chill and garnish each bowl with some of the fromage frais or yogurt and a reserved watercress sprig.

If you want to serve the soup hot, heat it carefully ensuring that it does not boil. Remove it from the heat and garnish as before.

DOROTHY HARTLEY'S WATERCRESS SALAD

Dice some cooked waxy new potatoes and drop them into a very little salted cream. Cover the bottom of a flat dish with some sliced tomatoes. Take a small bunch of parsley and shred this finely over the tomatoes. On this green bed, lay a layer of the white potatoes, and finally cover with watercress sprigs. Serve dry, and hand oil and vinegar separately. Do not use pepper with this salad.

WATERCRESS MOULDS

Made using individual ramekins, these make an excellent and very attractive first course.

1 bunch of watercress, stems removed
2 eggs
150 ml / 5 fl oz sour cream
sea salt
radicchio leaves, to garnish
for the green peppercorn sauce:
150 ml / 5 fl oz sour cream
2 tbsp green peppercorns in brine

Preheat the oven to 190°C/375°F/gas5.

Reserving a few whole sprigs for garnish, place the watercress in a liquidizer with the eggs and blend. Mix in the sour cream and salt.

Butter four ramekins and pour in the mixture. Place the ramekins in a bain-marie and bake for 15 minutes, or until risen.

While they are baking, make the green peppercorn sauce by mixing the two ingredients together thoroughly.

Take the moulds out of the oven and leave to cool slightly, then unmould on to individual plates. Garnish with radicchio and the reserved watercress.

Spoon a little of the green pepper sauce, not more than a teaspoon or two, over the top of the moulds.

LAND CRESS

Barbarea verna

What a pity this excellent green salad plant with very pretty edible yellow flowers is not easily available in farm shops and supermarkets. It could otherwise become a big commercial crop as it is very easy to grow and, in flavour has much of the peppery delight that we find in watercress. However, we lucky people who have gardens can cultivate it. If sewn in late summer and given the protection of a cloche or cold frame (so that severe frosts will not damage it), land cress can be picked throughout the winter. Its nutritional value is similar to that of watercress.

GARDEN CRESS

Lepidium sativum

This is the cress which is traditionally grown with mustard, or it may sometimes be rape (*Brassica napus*). The seedlings of either of the latter germinate 3–4 days earlier than cress. This is why punnets of both might initially only contain the mustard.

Cress was originally a native of Persia, from whence it spread into the gardens of India, Syria, Greece and Egypt. Xenophon had a theory that the Persians ate cress before they baked bread. Other than the common cress we know, there are three different kinds – curled, broad-leaved and golden. All were familiar in the Ancient World. In fact, it is likely that they knew others. Pliny talks of a giant cress grown in Arabia. It is also interesting to discover how often bread is mentioned in connection with cress. Take Gerard: 'Galen saith that the Cresses may be eaten with bread and so the Ancient Spartans usually did; and the low-countries men many times do, who commonly use to feed of Cresses with bread and butter. It is eaten with other sallade herbes, as Tarragon and Rocket; and for this cause it is chiefly sown.'

Indeed, cress is still marvellous in sandwiches, excellent with a smear of mayonnaise, or a little ripe avocado or tomato or cucumber. Yet I see it around infrequently and perhaps we are not using it as we once did. It is easy to grow yourself, especially if you are into sprouting seeds and already have a sprouter. Remember, all seeds are a stored powerhouse of food and, once they start to sprout, their enzymes become active. Weight for weight, they provide more nutrients than any other natural food we know.

RADISH

Raphanus sativus

It is odd to think that the small radish which we might crunch with an aperitif has such an ancient history that its wild ancestor has entirely disappeared. The *raphanus* refers to its redness, possibly from the Sanskrit *rudhira* meaning blood. In antiquity it was cultivated all over the civilized world, so many different varieties were established at those earlier times from the Mediterranean to the Orient. It was then grown as a staple food plant, the leaves almost certainly being eaten as well as the root. Theophrastus lists five varieties and declares that the Boetian type was the best and the sweetest. This seems to be our salad radish: small and round, coloured red and white or reddish purple, with a leaf that can also be eaten as a salad when young.

Prior to the introduction of the olive tree, the radish was grown in Egypt for the oil from its seeds. In Italy, Pliny tells us that 'medical men recommend giving raw radishes with salt for the purpose of concentrating the crude humours of the bowels...' He goes on to tell us: 'They have a remarkable power of causing flatulence and eructation; consequently they are a vulgar article of diet, at all events if cabbage is eaten immediately after them, though if the radish itself is eaten with half-ripe olives, the eructation caused is less frequent and less offensive.'

The radishes of Egypt and Rome may have been of the winter radish group, which includes black radishes and the one we now know from Japanese cuisine, the daikon (see page 118). When the Greeks made their vegetable sacrifices to Apollo at Delphi, they presented turnips on a platter of lead, beetroots on one

ABOVE Varieties of small red radish with their leaves

of silver and radishes on gold plates, showing there was no doubt as to the esteem in which they held this particular vegetable.

The Romans brought radishes to England. Pliny observed that they grew better in the colder climes. The larger types of radishes were very likely the ones mentioned in accounts of medieval food. Albertus Magnus in the thirteenth century talks of a very large root of a pyramidal shape; this was grown all over northern Europe and was reported by the botanist Henry Lyte (1529–1607) as being the common radish of England. Gerard mentions four varieties as being grown in 1597 'eaten raw with bread' but, for the most part, 'used as a sauce with meates to procure appetite.' One supposes that the large white radishes were boiled and then puréed as a sauce for meats.

However, the small red variety was certainly eaten and enjoyed as a salad vegetable. Lady Fettiplace notes that 'radish prove best that sowen a day after the full moon in August.' Salads then, and for the next century, sound very familiar to us. The selection from a range of leaves is wide: lettuce, purslane, corn salad, sorrel, dandelion, mustard, cresses, the young leaves of radish, turnips and spinach, as well as the thinnings from rows of seedlings.

The last mention of the large white radish in English cooking appears, I believe, as late as 1845, when Eliza Acton gives one recipe for boiled turnip-radishes. She recommends leaving them whole and boiling them for up to 30 minutes, then sending them to table with melted butter or white sauce. She also adds that common radishes – when young, tied in bunches and boiled from 18 to 25 minutes – are very good on toast like asparagus.

Perhaps E. S. Dallas (who wrote *Kettner's Book of the Table* in 1877) should have the last word: 'There are few combinations of colour so beautiful and rich as the red and white of radishes against the green of the leaves. In glass dishes upon a dinner-table they are an ornament which may vie with the finest flowers.'

NUTRITION

Radishes are popular in slimming diets as they are so low in calories and protein. They are mostly water, but have a high vitamin B and C content as well as many minerals, including calcium and sodium. As in the onion family, the peppery flavour is caused by sulphur compounds.

Radishes have medicinal qualities; they were thought to be a remedy for stone, gravel and scorbutic conditions. They have a mild diuretic effect and relieve constipation and catarrh.

BUYING AND PREPARING

Choose crisp radishes with springy bright green leaves – an excellent sign of freshness, as the leaves wilt very quickly.

The small red radish is best eaten raw, whole or sliced in salads. However, they tend to get lost if just added to a green salad. Try dishes where they are diced, marinated in a vinaigrette and mixed with dried and fresh fruit, with a sprinkling of roasted pine nuts perhaps. Explore their many possibilities by varying the salad recipe below.

RADISH, ORANGE AND PINE NUT SALAD

55 g / 2 oz dried apricots, sliced
about 150 ml / ¼ pt dry sherry
1 tbsp sesame oil
2 tbsp olive oil
1 tbsp white wine vinegar
1 bunch of radishes, trimmed and washed
2 oranges, peeled and sliced
30 g / 1 oz pine nuts
2 tbsp hazelnut oil
1 slice of wholemeal bread, cubed
2 garlic cloves
sea salt and freshly ground black pepper

The day before: in a small bowl, soak the slices of dried apricot overnight in just enough of the dry sherry to cover them.

Make a dressing by mixing the sesame and olive oils with the white wine vinegar and seasonings. Slice or quarter the radishes and add them with the apricots and any of their liquid. Leave them to marinate for about half an hour.

Peel and slice the oranges, catching the juice, and add both to the radishes.

Fry the pine nuts in a little hazelnut oil until just brown, then fry the bread with the garlic until golden in colour. Sprinkle these over the salad.

LARGE WHITE RADISH or DAIKON or MOOLI

Raphanus sativus spp.

'The radish occupies a far more exalted position in Oriental cultures than it does in the West, and is much larger than its Western counterpart. It is the most widely grown vegetable in Japan, where, so I've been told, gift-wrapped radishes are a token of esteem.' So says Joy Larkcom, giving ten pages of detailed consideration to the various radishes used in the Orient. They are used, as she says, either for cooking or pickling, a few are used for salads, others are salted and dried to preserve them. 'Radish leaves, stems, seed pods and seedlings are all valued as vegetables, cooked, pickled or raw.'

Whenever I have bought them, the large white radishes we see on sale here have turned out to be so mild they are almost tasteless. I suspect that, as they keep well in appearance, they are kept for too long and they sadly lose their flavour. The ones I have eaten from my own garden, on the other hand, have been verging upon the hot and peppery. These are a welcome addition to the stir-fry. I have also grated these and eaten them raw in a vinaigrette with the addition of carrot or grated apple.

For stir-frying white radish, the vegetable can be sliced across in discs or down in slices. It really

RIGHT *Black radishes, large white radishes and assorted small red radishes*

depends on whatever vegetable you are using and how you want the finished dish to appear.

As a cooked vegetable, it can be trimmed and cut and poached in vegetable or chicken stock for 5 minutes. The cooking liquor is then thickened with a little potato flour and some chopped parsley is added before serving.

BLACK RADISH

Raphanus nigra

This does not seem to have been mentioned by the writers of antiquity. It is sometimes known as the long-rooted Black Spanish, and there is also a round variety. They can easily be grown in Britain, but must be picked young, for they quickly go woody and are inedible. Boiled with their skins on for about 10 minutes, then peeled and sliced, they are excellent with an *aïoli* mayonnaise.

PICKLED SEED PODS

There are numerous recipes throughout history for the pickling of the radish seed pods. Indeed, in India and China the green pods are pickled and are used in stir-fried dishes. Here is Dorothy Hartley's recipe: 'Pick the pods when the seeds inside are like soft seed pearls, drop the pods into boiling brine, and let it grow cold – if they are now bright green you can pickle at once, but if dull, boil up the brine and pour it back over the pods once or twice till they are bright as emeralds. Drain, rinse clear of salt, pack the pods into glass bottles, with one or two scarlet peppers, cover with white clear vinegar, and cork down for winter use. They are a very pretty garnish to winter's potato salad.

The clear green pods in a dish of pellucid pickled onions make a jade and pearl symphony that belies its potency.They are also delicious (and most decorative) with cream cheese.'

HORSERADISH

Cochlearia armoracia

The horseradish plant was indigenous to eastern Europe, Russia, Poland and Finland and is thought to have been brought down through the Caspian to Persia by the Aryans. The Egyptians used it as a medicine and a flavouring, It was popular in Rome and referred to by Dioscorides. Both the leaves and roots were eaten in Germany during the Middle Ages, but it does not appear to have come to Britain until later. Gerard mentions that the Germans use it and the botanist and antiquary Lyte in 1586 talks of the wild horseradish being used as a condiment.

It began to be popular here in the seventeenth century, when the oilier fish like mackerel, herrings and salmon were served with horseradish grated over them. As the stronger pickles were being introduced by the East India trade, so a liking for hotter sauces developed. Many pickles were now manufactured in Britain and the pickling liquor almost inevitably included horseradish.

Horseradish sauce began to be eaten with roast beef in the nineteenth century. A recipe of the time suggests grating the root finely, adding a gill of cream, a dessertspoonful of sugar, a little salt and rather more than a tablespoonful of vinegar, and mixing well together. Another Victorian recipe from *Kettner's Book of the Table* suggests adding the grated radish to the zest and juice of an orange, three tablespoons of oil, a tablespoon of breadcrumbs and one of vinegar with a teaspoon of sugar and a pinch of salt.

Few of us have ever seen the horseradish root unless we grow it. Once grown in a garden it tends to take over and it is difficult to eradicate, for new plants will grow up from the smallest sliver of root left in the ground. If you are lucky enough to find a root, scrub and clean it well, for the skin and the outer part is the hottest and most aromatic. Grate the outside, trying to protect eyes and nostrils, for the volatile oils are of the pungent mustard kind. The best method is to use a food processor, which will give some protection from the fumes. Combine it with cream or yogurt or smetana, and other ingredi-

ABOVE A Japanese horseradish farm

ents as per the previous paragraph. Grated horseradish soon loses its pungency, which is why the bottled type is always so disappointing.

There are several other varieties of horseradish used as food in other parts of the world. In the far north and Arctic regions, the leaves of *C. danica* are eaten as a salad. In Hungary and the Balkan peninsula, they use the root of *C. macrocarpa* grated, but this variety is somewhat less acrid, while *C. officinalis* was eaten as a cress in salads in Scotland and farther north. Commonly called 'scurvy grass' or spoonwort, it was also used on long sea voyages as a preventative to scurvy. They even made scurvy-grass ale which became a popular tonic drink, although the name was somewhat of a misnomer as the grass was not fermented. It was, in fact, a tisane: some of the leaves were infused in boiling water and it was taken in wineglassful doses throughout the day.

JAPANESE HORSERADISH

Wasabia japonica

Though it is not closely related to our horseradish, this is still a member of the cabbage family. It has grown wild on the Japanese coast and up in the mountains, and likes wet ground beside streams. It is used fresh as its pungency declines quickly and is a bright green in colour. It is generally grated to serve with *sashimi*, assorted finely sliced raw fish

We can only obtain it in the West after it has been dried and pulverized into a powder. Unlike grated Western horseradish, powdered wasabi kept in its tin retains almost all of its pungency. Make some up by mixing it with water and then leave it for 10 – 15 minutes, for only then will the real fiery intensity of the root be made manifest.

ROCKET

Eruca sativa

Of all the salad plants and their leaves, this is my own favourite by far. When small, the plant has green leaves which are slightly reminiscent of radish, dandelion or an oak-leaved lettuce, but the taste is singular – peppery and growing more mustard-like as the plant ages. The leaves bequeath enormous zest to a mixed salad, though in my opinion the flavour is better without dressing. I use it extensively as part of a first course, its striking flavour a marvellous foil as a base for grilled chèvre, sliced avocado, or a variety of purées and salads.

Yet there is a great puzzle surrounding this edible and almost addictive plant. Why did it disappear entirely from Britain's gardens and thence our table for almost three hundred years? Further, one might ask, does the belief that the leaves have erotic properties have anything to do with that disappearance?

There is no doubt that the Ancient World considered it as possessing aphrodisiac properties; it was sown around the base of statues consecrated to Priapus and believed to restore vigour to the genitalia. Ovid terms the herb 'salacious', while Martial links it with spring onions as 'lustful'. Apicius uses both the seeds and the leaves, grinding the first as an ingredient in aromatic salts to be added as flavouring (it appears in a sauce for cold boar), while the latter is part of a sauce for boiled crane and the pounded leaves with other herbs are used as a dressing for salt grey mullet. In one recipe Apicius acknowledges rocket's salacious reputation and teams it with bulbs 'for those who seek the door of love, or as they are served with a legitimate wedding meal, but also with pine kernels or flavoured with rocket and pepper.'

The plant has a long history, for in the Middle East between 9000 and 7000 BC rocket was one of the first plants to be cultivated. It originated as a weed with both rye and oats, infiltrating the very first cultivated plants, wheat, barley, rice, soya beans, flax and cotton. As cultivation stretched upwards from the

RIGHT Rocket and seakale

river basin to the higher regions, or travelled north to harsher climates, the so-called weeds were better adapted to these conditions and so took over the fields. Thus the weeds became the crop and the crop the weed. (In the same way, the tomato plant was a weed which spread northwards from Peru through the New World tropics to Mexico as a weed of both maize and bean fields.)

Unaware that rocket was one of the first cultivated plants, the early Christian church only knew of its supposed erotic properties and frowned upon its use and cultivation. Numerous writers spoke of its 'hotness and lechery' and the Church, at one point, banned its cultivation in monastic gardens.

Its common name is derived from the Latin '*eruca*', which denotes a downy stem like a caterpillar. Giacomo Castelvetro in 1614 delights in its use as part of an early spring salad (see below). John Evelyn's rocket is derived from Spain to be planted in his calendar in March. In the late 1690s Evelyn refers to it as both 'hot and dry', but 50 years later in 1747 Hannah Glasse omits rocket from her list of salads. Had it, by then, fallen out of favour? It would seem so.

Its decline is traced quite clearly in Sturtevant's *Notes on Edible Plants* (published 1919). Here he says, 'in 1586 Camerarius says it is planted most abundantly in gardens. In 1726 Townsend says it is not now very common in English gardens, and in 1807 Miller's Dictionary says it has been long rejected.' It is interesting to note that this decline was not paralleled elsewhere; it continued to flourish in gardens and cooking all over the Mediterranean countries and it appeared in American gardens in 1854 or earlier.

John Evelyn shared the seventeenth-century passion for novelty in gardens; new varieties of fruit and vegetables were grown with great enthusiasm, yet were they eaten with the same enthusiasm? One doubts it. Pepys hardly mentions vegetables, except as ingredients for broths. For people were suspicious of vegetables, believing them (quite rightly) to be a source of unwanted wind. They were also thought to cause melancholy. Though it is doubtful that rocket was ever eaten here cooked (as it was and still is in the Mediterranean), but as part of a salad it would have been well-seasoned with oil as this was thought an antidote to flatulence. All raw, undressed food was much frowned upon.

What could possibly be the reason, then, for the sudden disappearance of rocket from the British salad bowl, kitchen and herb garden? After Evelyn it stops being eaten in these islands, until it reappears in seed catalogues around 1980. Within this short time it has risen astonishingly quickly in popularity, so that every enterprising gardener and cook grows it, and the large multiple stores stock it.

It would seem that the disappearance of rocket and other salad vegetables in this country was due to the rise in market gardens. The rapid growth and prosperity of towns favoured the development of market gardens situated nearby. The vegetables and fruits grown in these gardens were chosen from the stock in the great houses and they tended to be the most popular, which could be sold off easily: peas, beans, potatoes, carrots, celery, cabbages, cauliflowers and Savoys and the root vegetables, turnips and parsnips. Some of these vegetables became field crops; others gradually came to be grown, but not until a hundred years later, as part of the garden produce around the house to eke out the family diet. The rule of selection favoured what could be cultivated and sold easily. This would seem to apply to rocket, at first glance, as this salad plant seems too wild in appearance and habit to fit into the well-ordered market garden. Rocket can, after all, be found wild and other salad leaves could certainly have been picked easily from the countryside.

A countryman with a small plot of land early in the nineteenth century could grow his own vegetables. Cobbett at Singleton in Sussex in 1823 notes, 'the gardens are neat and full of vegetables of the best kinds. I see very few of "Ireland's lazy root" (meaning the potato)'. The vegetables all tend to be the ones mentioned above with emphasis upon the brassicas. By this time, if rocket had strayed into such a garden it would have been plucked out as a weed. This conservatism is still rife today in the rural gardener, who looks upon my own garden of salad leaves, radicchio,

RIGHT *Rocket, mushroom and flageolet salad (page 128)*

endive, chicory and mustard greens as being alien, and filled with plants not worth giving room to in his own plot of cabbages, beetroots, leeks and onions.

The evidence for rocket's downfall and obscurity is even more striking if one consults *A Modern Herbal* by Mrs Grieve, published in 1931 with an introduction by Mrs Leyel, and long considered a modern classic. Here there is an entry for 'Rocket, Garden', which refers to *Hesperis matronalis*, sweet rocket, but which then lists *Eruca sativa* as a synonym. It also goes on to make another profound mistake by saying that the rocket found in the wild in Britain and Russian Asia are escapes from gardens. Mrs Grieve gives no separate entry for the great culinary rocket, *E. sativa*, an omission which is very odd in such a comprehensive work some 900 pages long. It is even more astonishing when one realizes that a small volume by Lady Rosalind Northcote, *The Book of Herbs*, published in 1912, gives a nervous appraisal of the herb, but refers back to the 1629 work of Parkinson (herbalist and apothecary to James I), *Paradisi in Sole*, who believed rocket 'causeth headache and heateth too much'. Lady Northcote ends her half-page reference by commenting 'it gives little encouragement to those who would make trial of rocket'.

Well, no, she is right. What a pity both Mrs Grieve and Lady Northcote did not take more notice of John Evelyn. Could it be that the ancients' love and admiration for rocket as an aphrodisiac made later writers nervous of it? Especially those who might be influenced by the severe admonitions of the medieval church, such as Edwardian gentlewomen like Mrs Grieve, who cast it completely out of her mind and from thence the botanical world entirely.

CULTIVATION

Rocket is enormously easy to grow. Get into the habit of successive sowings from early March, depending on demand (one every two months is my own usual pattern), but be sure to thin the seedlings and use them in a salad as soon as possible, as well as cutting back the leaves as they are growing. Rocket is a sturdy and flourishing plant which goes to seed easily if it is not cut back and eaten. The small white and yellow flowers can also be used in salads or as garnish. The plants will seed themselves, or the seed can be collected and, once ground, can also be used as flavouring. If you have a cold greenhouse you can keep rocket going throughout the winter months. Outside, the mature plants will often survive snow and frost.

SWEET ROCKET

Hesperis matronalis

This is a biennial which flowers early in March, in either white or lilac, and continues flowering for several months. I keep a few plants going in my herb garden, for it is not only charming but the young leaves can also be eaten in salads. They are very high in vitamin C and are valued medicinally.

YELLOW ROCKET or WINTER CRESS

Barbarea vulgaris

This is another biennial which grows wild, not to be confused with land cress (see page 116), though the flavour is similar. They can also be boiled or stir-fried. It is a sturdy plant which will flourish the more the leaves and yellow flowers are harvested.

COOKING

Rocket is seldom cooked here, though it is always part of the wild greens called *horta* which are gathered and cooked in Greece in the autumn. They are simply boiled for 5 minutes, then well drained and served with oil and lemon. Patience Gray uses rocket with pasta in *Honey from the Weed*:

'Orecchiette con la Rucola' – "Little ears" with rocket (Orecchiette are the traditional pasta of Apulia, shaped like little shells, particularly useful for holding the sauce.)

A Salentine dish. Use wild rocket, *Eruca sativa* (or broccoli heads or heads of rape). Gather the plants when small, wash them well and throw into a pan of salted boiling water. Cook the orecchiette in another pan and drain when they are al dente. Cover the bottom of a frying-pan with olive oil, add 2 hot peppers, 2 peeled cloves of garlic, sliced, and cook for a few minutes. Put the well-drained pasta and the rocket in the pan, stir with a wooden fork, mixing all together,

ABOVE *Rocket and avocado sandwich (overleaf)*

and serve very hot with a piquant grated cheese.'

However, rocket is best when used in a salad or by itself, perhaps Castelvetro's 'excellent mixed salad' is the most beguiling of all.

'Of all the salads we eat in the spring, the mixed salad is the best and most wonderful of all. Take young leaves of mint, those of garden cress, basil, lemon balm, the tips of salad burnet, tarragon, the flowers and tenderest leaves of borage, the flowers of swine cress, the young shoots of fennel, leaves of rocket, of sorrel, rosemary flowers, some sweet violets, and the tenderest leaves or the hearts of lettuce. When these precious herbs have been picked clean and washed in several waters, and dried a little with a clean linen cloth they are dressed as usual, with oil, salt and vinegar.'

No writer that I have yet to discover mentions that the wonderful peppery flavour of rocket will be diminished once it is dressed with oil and vinegar. If you want to retain that crisp taste, almost of green peppercorns, then keep some of the rocket leaves undressed. This is my method: lay the rocket leaves on a platter, then mix the other salad ingredients with the dressing and pile that on top of the rocket in the centre. That ensures that some leaves around the edge remain undressed.

ROCKET, MUSHROOM AND FLAGEOLET SALAD

generous handful of rocket leaves
4 or 5 spring onions, chopped
5 or 6 chestnut mushrooms, sliced
2 or 3 tbsp cooked flageolet beans
4 tbsp walnut oil
1 tbsp lemon juice
pinch of caster sugar
sea salt
shreds of lemon zest, to garnish (optional)

Lay the rocket leaves over a serving platter.

In a bowl, mix all the other ingredients with salt to taste and let them marinate for an hour.

Then pile the mixture in the centre of the rocket-lined platter and serve garnished with some strips of lemon zest, if using.

ROCKET AND AVOCADO SANDWICH

This is my favourite summer sandwich. Rocket and avocado go marvellously well together, but I confess I like a touch of other flavourings in it as well, as a kind of background track to the two stars.

This amount will make two fairly large and bulky sandwiches, which provide an excellent and satisfying lunch for two people.

2 tbsp soft goats' cheese
4 slices of your favourite brown bread, buttered
smear of Marmite, Vegemite or Vecon
about 10 thin cucumber slices
1 ripe avocado, peeled, stoned and sliced
20 or so rocket leaves
tiny drop of Tabasco (optional)
sea salt and freshly ground black pepper

Smear the goats' cheese on the buttered side of 2 slices of the bread and smear the Marmite or whatever on the other 2 slices.

Lay the cucumber slices over the cheese, season with pepper, then lay the avocado slices on top of that, followed by the rocket leaves. Season with salt and Tabasco, if using.

Then cover with the other pieces of the brown bread. Press the tops gently down and slice the sandwiches in half with great care.

SEAKALE

Crambe maritima

This is a beautiful and mysterious plant when still in its wild state and, when cultivated (I have several plants in my own garden), it looks no different from the seakale you will find on the beach. There it sits like a great silver-green bouquet splayed over the shingle, astonishing to consider that the shoots have thrust their way through heavy pebbles, gravel and sand towards the sunlight.

It is the stems of the potential leaves, the shoots, which are blanched the minute they appear, by piling seaweed or gravel on them; or, in cultivation, by a blanching pot which is then covered in manure so that it is warm, to hurry growth.

Seakale grows around the sandy shores of the North Sea, the Atlantic Ocean and the Mediterranean Sea. The Romans gathered it and preserved it in barrels for eating through long sea voyages. It is recorded as being brought to English culture from Italy about 1760 and the seed then sold at a high price as a rarity. However, sea folk must have eaten the plant since time began, knowing the places along the shore where it grows and waiting to cover the growing shoots with seaweed to blanch them.

It was Louis XIV's gardener, la Quintinie, who first cultivated it in France. The people of Sussex were gathering seakale in the spring and selling bundles of it at Chichester market; there is a record of this dated 1753. Thus the seed from Italy which sold at such a high price must have come here through gourmet demand. So much so that by 1799 the botanist William Curtis wrote a booklet on the culture of seakale for the table and for a time it was even cultivated commercially. It remained a popular vegetable with the elite up to about the end of the nineteenth century.

Carême discovered it, to his great delight, on sale in London: 'They resemble branches of celery and are to be served like asparagus with a butter sauce, but I prefer to serve them with Espagnole.' (Espagnole sauce is a classic brown sauce which has had some slices of Serrano ham cooked in it.) He advised as much as a lengthy 20 minutes' boiling.

If it was this popular, why has seakale all but disappeared now? It must be yet another casualty of our growing urbanization, so that we have cut ourselves off from wild plants which still thrive in uninhabited areas and near desolate beaches. Also, to harvest seakale in the spring, you need to know where the plants are. For this they have to be marked the summer before. Or, like me, you can simply grow it in your back garden.

The thongs with terminal buds for planting can be purchased by mail order (see the list of suppliers at the end of the book). Alas, you cannot harvest the shoots for two years, but then, what bliss. Simply boil the shoots for 10 minutes and then eat them, like asparagus, with a little melted butter.

BELOW *Seakale growing in a garden*

Cucurbitaceae

THE SQUASH FAMILY

SQUASHES and PUMPKINS
Cucurbita pepo and *C. maxima*
VEGETABLE MARROW and COURGETTE
Cucurbita pepo
VEGETABLE SPAGHETTI *Cucurbita pepo*
CHAYOTE *Sechium edule*
CUCUMBER *Cucumis sativus*
GHERKIN *Cucumis sativus* and *C. anguria*
CHINESE BITTER MELON *Momordica charantia*
SMOOTH and ANGLED LOOFAHS
Luffa cylindrica and *L. acutangula*
BOTTLE GOURD *Lagenaria siceraria*

Most of this family's 805 species climb by tendrils or have a sprawling, prostrate habit. They are perhaps most striking for their Triffid-like pullulation of growth. Nothing else in the garden, except weeds, grows at such speed. Leaves, tendrils, stalks and flowers expand with rapidity, covering the soil beneath with their hairy lime-green leaves and yellow flowers the colour of thick cream. They are avaricious for space, having a tenacity – almost an aggression – in their expansion. Some gardeners make use of this high growth rate by growing them up arches or pergolas, using the plants decoratively.

The fruits also expand at a very fast rate. Melons at their most active put on 80 cc / 5 cu in a day; pumpkins can average about 340 g / 12 oz, and daily gains of almost twice that have been known. This expansion is due to the accumulation of water-based sap in the cell vacuoles. The high water content of the juicy flesh makes the fruits particularly refreshing in hot weather and tropical climes (it also makes some of them insipid when water is added during cooking). The rest is seeds, the whole protected by a more or less hard skin.

Annuals native to temperate and tropical areas, they tend to be sensitive to temperatures near freezing, which limits where they grow and how they are cultivated. Cucumbers, melons and gourds are Old World plants known to have been under cultivation for 4,000 years; pumpkins and the other squashes all stem from the New World and some of the 25 species (probably the long-lasting winter squashes, with their oil-rich seeds) have been grown for 9,000 years.

SQUASHES and PUMPKINS

Marrows, squashes and pumpkins all belong to the gourd section of the family – all are varieties of *Cucurbita pepo* or *C. maxima* – and come from the New World. It is a confusing family for the enterprising shopper, interested in trying unfamiliar vegetables, for the fruits come in all shapes, colours and sizes, and carry a multiplicity of names. Sometimes the fruits from an identical plant are known by different names according to whether they are harvested small or mature (as with courgette and marrow). Also the same variety is known by different names in different countries. Increasingly the squashes found in supermarkets bear their North American names, while the same ones in ethnic stores bear their Oriental or Caribbean names.

The whole multifarious squash family was, of course, part of the riches discovered in the Americas and described by the early explorers. They were unknown to the Old World before the time of Columbus, despite some confusion on this topic. Jane Grigson in her marvellous *Vegetable Book* is one of those who fall into the easy trap of believing that pumpkins were known in the Ancient World. Waverley Root, exploring this conundrum, points out that squashes are supposed to have grown in the gardens of Babylon, and that Pliny, Apicius and Martial all mention them – in fact the last gave a dinner composed entirely of different kinds of squash. But the Latin word could also be translated as 'gourd': Albertus Magnus mentions edible gourds, and the French '*citrouille*' can mean either pumpkin or gourd. What was being eaten in the Old World was fruit from the *Cucumis* section of the Cucurbitaceae (cucumbers, melons and the edible gourds that are still preferred in both China and India today as being far more tasty than squashes). This is endorsed by the fact that there is no word for squash in Sanscrit, and none is mentioned either in the Bible or in any of the ancient Chinese writings. No trace of squash or pumpkin has been found in Egyptian tombs – only water flasks made from gourds.

The confusion originated in the familiar way, when people applied an old name to a new thing. Our word squash is obviously borrowed from the American Indian name '*askutasquash*', but many of the early explorers after Columbus were constantly calling the squashes 'gourds'. In 1672 John Josselyn, the English author of *New England Rarities discovered*, tries to describe the new fruit: 'A kind of melon or rather gourd, for they sometimes degenerate into gourds. Some of them are green, some yellow, some longish, like a gourd, others round like an apple...' He adds, 'all of them are pleasant food, boyled and buttered and seasoned with spice.' To add to the confusion, the pattern was repeated in Spanish, and even today in Cuba large squashes and pumpkins are called '*calabazas*' from the Spanish for 'gourd'.

One useful division is into summer and winter squashes. Essentially, summer squashes are good fresh and do not store long. Most are fairly thin-skinned. They are edible raw when young: try them grated and salted and pressed into moulds (see page 138). If picked young they can be eaten unpeeled, so they retain more flavour and nutritional quality. They are excellent cut into strips and stir-fried. Their flavour is light, quite bland but always refreshing – an excellent foil to stronger-tasting vegetables and to spices.

ABOVE An assortment of squashes and pumpkins

At a blind tasting, only the greatest of squash *aficionados* could detect any difference in flavour and tell which variety was which. Cooking methods are virtually interchangeable: use any recipe for courgette, marrow or squash, and try recipes for pumpkin, too.

Winter squashes, on the other hand, can be eaten when young but have the advantage of keeping well when allowed to ripen at a good temperature and stored in a frost-free place. Most are larger and thicker-skinned than summer squashes, and many have hard, inedible seeds. *Curcurbita maxima* varieties take twice as long to grow as summer squashes – needing about 110 days to reach maturity – and are rarely grown in countries that do not enjoy long, hot summers, unless they are started under glass. The texture is firmer and more floury than that of summer squashes. But they do contain more protein, fat, carbohydrates and considerably more vitamin A than summer squashes such as marrows. They are excellent additions to winter soups, thickening and flavouring a mixed vegetable selection. They are also good for making preserves and home-made wine.

PUMPKIN

Cucurbita pepo and *C. maxima*

The pumpkin is perhaps the most famous winter squash with its deep orange flesh and its traditional role as part of American Thanksgiving. Pumpkins seemed an obliging fruit. They grew in the south and in the north of the Americas. Pedro de Alvarado, reporting in 1540 on Coronado's penetration of the south-west, says that the territory grows melons. In 1584 Jacques Cartier reports from the St Lawrence region that he has found there '*gros melons*', a phrase translated into English not as 'big melons' but as

ABOVE Ginger and pumpkin soup and courgette moulds (see page 138) served with pumpkin sauce

'pompions'. Hence our word 'pumpkins'. The English translation of Estienne and Liébault's *Maison Rustique* includes the advice: 'To make pompions keep long, and not spoiled or rotted, you must sprinkle them with the juice of a houseleek...' (Houseleek or common stonecrop used to grow on walls and was thought to have considerable medicinal powers.)

Pumpkins can grow to a considerable size. Pumpkins weighing two hundredweight – 100 kg or 224 lb – were a common sight in London markets of the nineteenth century. No wonder they seemed like a possible starting point for Cinderella's coach. From the seventeenth century the flesh was used, mashed and puréed, to bulk out bread.

The pumpkin's great quality is its colour – that fiery russet or Van Gogh orange sets the table aflame. The fact that its flesh melts easily into a velvet-smooth purée makes it highly suitable for soups and sauces. Its largeness also invites use as a container, hot or iced, for casseroles or salads. Try the Iced Pumpkin Salad (see page 136), a great luncheon dish for an Indian summer.

NUTRITION

The pumpkin is fairly rich in vitamin A, folic acid and potassium, with small amounts of the B vitamins, fibre and iron.

CHOOSING AND STORING

Make sure a pumpkin is firm, without bruising. They come in all sizes and weights, so buy only what you need. Often a market stall or store will have pre-sliced pumpkin or be ready to cut off the portion you need. As a general rule the more intense in colour the orange flesh, the more vitamins and taste it has. Also, the larger the pumpkin, the less flavour the flesh will have. So go for small, squat pumpkins with a tough green rind and bright orange flesh. Cut pumpkin wrapped in clingfilm will keep in the refrigerator for a week to ten days.

COOKING

Pumpkin can be used in many ways. Cubes of flesh can be fried in oil and garlic and served with fried pine nuts and chopped leaf coriander. Cubes can also be put in a pan with a little butter and oil and left over a low heat to steam in their own juice; they will cook inside 15 minutes. The pumpkin can then be seasoned and puréed to use as a sauce or a basis for soup. Cubed pumpkin can also be the beginning of a risotto, a Lombardy dish.

If you want to cook the whole pumpkin using the shell as a casserole dish (a customary recipe to serve for a Hallowe'en party), it is best to start the day before. This gives time for all the flavours to amalgamate and the actual cooking inside the pumpkin (see recipe) can be done on the night. You will need a pumpkin weighing about 4.5 kg / 10 lb for this, but (as Jane Grigson points out), do measure the inside of your oven to see what size of pumpkin will fit.

Chunks of pumpkin can also be roasted in a pan around a joint or bird. This is particularly good with game, which is in season at the same time. Cut large chunks 10 cm / 4 in square, remove the peel and simply let them brown in the fat, turning them over as you would any other roasting vegetable.

Pumpkin seeds can be dried and eaten, as they were on the table of Montezuma.

GINGER AND PUMPKIN SOUP

55 g / 2 oz root ginger, peeled and grated
1 tbsp olive oil
675 g / 1½ lb pumpkin flesh, cubed
30 g / 1 oz butter
575 ml – 1.1 litre / 1 – 2 pt vegetable stock

This is simplicity itself. Throw the ginger into a large pan with the oil and butter, sauté for a few seconds, then add the pumpkin. Place a close-fitting lid on the pan and leave over a low heat for 15 minutes.

Leave to cool and then liquidize, adding enough of the stock to make the consistency of soup that you wish. You will achieve a richer flavour by adding more butter, but the flavour of pumpkin is so excellent it needs, in my opinion, very little.

PUMPKIN SAUCE

Omit the ginger (if you wish) and cook as above. Leave out the stock completely, to make a thick, rich sauce; or thin down with a little stock for a runnier consistency. Wonderful to use as a puddle for first courses, as an alternative to tomato coulis.

BAKED PUMPKIN CASSEROLE

115 g / 4 oz flageolet beans, soaked for 1 hour in boiling water and drained
115 g / 4 oz haricot beans, prepared as above and drained
115 g / 4 oz porcini (ceps)
225 g / 8 oz field mushrooms
½ tsp asafoetida (optional)
1 tsp turmeric
2 large onions, sliced
2 green chillies, sliced
2 green peppers, cored, seeded and sliced
450 g / 1 lb tomatoes, peeled and chopped
900 g / 2 lb sweet potatoes, peeled and diced
4 heads of sweetcorn, cooked and scraped off the cob
850 ml / 1½ pt vegetable stock
3 tbsp olive oil
3 tbsp tomato purée
sea salt and freshly ground black pepper

Pick a large pumpkin which will fit in the oven. Cut the top off, complete with its stalk, about 8 cm / 3 in down and reserve it to serve as a lid. With a knife and a spoon take out all the seeds and cottony fibres and discard. Then, very carefully, scoop out some of the flesh from the inside, leaving a good layer of flesh 2.5 cm / 1 in thick adhering to the wall. This will be cooked and so will soften and disintegrate into the stew, but you need the outside walls to be strong enough not to collapse, so be cautious. It is best to cut away too little than too much, so go carefully to begin with; if you puncture the outside wall you will have to call it a day and start all over again with a new pumpkin.

Preheat the oven to 190°C/375°F/gas5 and with its lid back on place the pumpkin into the oven and bake for 20 – 30 minutes. It will need another 10 minutes cooking with the casserole inside it. This can be done immediately if you wish, in which case when you come to reheat the pumpkin, it will need another 15 minutes in the oven.

Cook the beans in separate saucepans; the flageolets should need only 20 – 30 minutes, while the haricots could need up to one hour. Boil them without salt, then drain and reserve.

Heat the olive oil in a large saucepan and throw in the porcini, mushrooms and spices, then the onions, chillies and peppers. Sauté all this for 2 – 3 minutes, then add the tomatoes, the scooped-out pumpkin flesh, sweet potatoes and 575 ml / 1 pt stock. Simmer for about 20 minutes, then add in the beans and sweetcorn, tomato purée and seasoning. If too thick, add the extra 300 ml / ½ pt of stock, but remember that the cooked pumpkin will contain some liquid, so it is best to reserve the stock until after the stew has been added.

Spoon the stew into the pumpkin and reheat in the oven for 10-15 minutes. The pumpkin may only take about half the stew – it depends on the size. Reheat the remainder in an ovenproof dish. Serve with good crusty bread.

Jane Grigson advises that if you have a large casserole into which the pumpkin can be settled, it is prudent to sit it in it to use it as a kind of support to the base and lower walls.

ICED PUMPKIN SALAD

This uses the same technique of cooking the whole pumpkin described above; after baking it, let it cool, then pour off any liquid (discard it or drink it). The pumpkin flesh you have scooped out is cubed and fried in olive oil and garlic until it is just golden brown. Reserve and cool.

Then lightly steam a collection of baby vegetables, such as small carrots, courgettes and leeks for about 3 minutes. They need to have plenty of bite in them. Chop some parsley, chives and basil. Mix in with the baby vegetables, add some baby mushrooms and tomatoes and a chopped yellow or red pepper, all raw. Make a thick *aïoli* mayonnaise and mix it with the vegetables and the cooked pumpkin. Pile it into the cooked whole pumpkin so that it just overflows and decorate with any herb garnish you have, placing the lid gently on before serving.

Dorothy Hartley in *Food in England* gives a version of this salad, but uses a vegetable marrow.

VEGETABLE MARROW and COURGETTE

Cucurbita pepo

The vegetable marrow is a summer squash that endures much disparagement from cookery writers – probably because most encounter it too late in the day, when its skin has become tough and its flesh watery and insipid, and when prepared in a way that has not done it justice. In its prime (after maturing from the courgette stage but before going downhill into old age), the marrow is perfectly capable of playing a role in the theatre of gourmet cuisine – a supporting role, perhaps – and of playing it beautifully.

Native to the Americas, with the rest of its family, it reached Britain in the nineteenth century. What made it popular as a cottage-garden vegetable was its ability to grow quickly and to fruit large within a very short growing season before the early frosts. The nineteenth century, contrary to ours, romanticized the marrow, thinking it came from Persia. E. S. Dallas in Kettner's *Book of the Table* waxes eloquently: 'Whoever brought them made a noble gift to his country. Of all the gourds in Europe the latest known, it is in England the most cultivated. It is, indeed, more prized in England than in any other European country, and can be obtained so cheaply that it is in great favour with the poor as well as with the rich. It is a watery vegetable without much nutriment, but it has a fine mellow flavour; and at the end of dinner, when we want something light to play with, its juicy slices make a delicious entremet.' He even gives a recipe – a Soup for the Shah – which in appearance was 'to represent the emeralds and topazes which the Persians love'. This recipe was astonishingly contrived with fried crusts, peas and asparagus points.

An *Encyclopaedia of Gardening* (1902) gives excellent advice on what it calls 'a cottager's vegetable': 'Usually the fruits are eaten whilst young and tender; if the skin is too hard to allow the entry of the thumbnail it is considered too old. Ripe Marrows make an excellent preserve.'

This is almost the last fair word the marrow gets. Mind you, the Victorians' liking was for marrow to be boiled and stuffed, but mostly boiled and for long periods. The vegetable hardly looks or tastes its best with this treatment, which merely aggravates its wateriness. Nevertheless, does it really deserve to be called 'this dreary vegetable', with 'its slimy, pappy taste' (by T. A. Layton in *Choose Your Vegetables*) or 'the Bunter of the kitchen garden' by Jane Grigson? I think not.

The golden rule, as all agree, is to follow the thumbnail rule and to pick the marrow when it is young, no longer than a foot. Once it grows larger, swelling up into a striped green balloon, it is best picked for the compost.

COURGETTE

This is essentially the marrow in its immature form, before the seeds and pith have formed inside and the skin has toughened enough to need removing. (The names we have adopted tell the story: 'courgette' is a diminutive of the French '*courge*', a marrow, and 'zucchini' means a miniature '*zucca*', Italian for gourd.) These juvenile marrows earn none of the opprobrium which their elders suffer. Very small courgettes (and particularly their flowers) are now part of high cuisine and deservedly so, though perhaps their appearance (rather than their flavour) is the main part of their attraction.

NUTRITION

One excellent quality the whole family possesses is that they are all low-calorie and a reasonable source of vitamins A and C; otherwise they contain carbohydrate, calcium and potassium.

VARIETIES

Marrow and courgette plants are virtually identical, though breeders have developed compact-growing marrow varieties with high cropping over a long period to give a regular supply of small fruits: the more you cut the more they produce; if you leave them on the plant they grow to marrow-like dimensions and flower production slows down.

There are marrows with green stripes (Minipac), which look attractive, and courgettes which are golden yellow (Gold Rush); Clarella is a green courgette with a delicate flavour, while Tondo di Nizza is a round courgette. A pale yellow (Bianco Friulano) has an attractive warty skin. Early varieties of courgette are Aristocrat, Diamond and Onyx.

CHOOSING AND STORING

Look for marrows with no yellow or brown patches. Use the thumbnail test if possible to make sure the skin is not too tough. They store well, for weeks if need be, if kept in the cool and the dark. If hung in string bags from hooks, they will keep all winter.

Choose courgettes that look quite fresh and firm, preferably no more than about 15 cm / 6 in long – even better if they are smaller.

In Italian markets you will see the flowers on sale, but they are delicate and easily droop and fade. If you grow courgettes, then the flowers can be picked and cooked within the hour. Pick the male flowers, borne on thin stems, and leave the female flowers on the plant so that their swollen stems can develop into more courgettes.

COOKING

Marrows can be peeled, their pith and seeds removed, then cut into cubes and fried in olive oil with garlic and ginger. Or cut into circles, dip in batter and fry. They are best, I believe – like yams, gourds and sweet potatoes – when used in Indian and Oriental recipes, for then their sweetness and delicacy of flavour can be offset by harsh or even aggressive spicing.

The smaller the courgette, the less cooking it needs. The really tiny ones, no bigger than one's thumb, merely need to be blanched or turned in butter, or both, before serving. Slightly larger courgettes can be sliced lengthwise, then fried quickly in olive oil or butter – a minute is quite long enough. Courgettes 15 cm / 6 in long need first to be trimmed, then perhaps sliced into eighths lengthwise for stir-frying, or cut across into discs for frying.

Larger courgettes can be poached or steamed. They can also be grated and stir-fried for a moment. The last method is very satisfactory, for they remain almost raw, yet lose about a third of their bulk in liquid which evaporates.

Grated courgettes can be salted and left in a colander to drain for an hour. Squeeze them dry, add chopped mint and chives and then press them into ramekins and refrigerate for a couple of hours. Serve these *Courgette Moulds* as a first course, turned out on individual plates with a few red salad leaves and sitting in a pool of either tomato coulis or *Pumpkin Sauce* (see page 135).

Called by their Italian name, zucchini are highly favoured in Italy. Anna del Conte mentions them blanched and covered with a creamy béchamel topped with Parmesan, sautéed in butter and oil and flavoured with oregano, or finished with tomato and basil. Courgettes are also stuffed; a recipe from Mantua uses onion, eggs, ricotta and Parmesan cheeses, stiffened with some crumbled amaretti biscuits.

Stuffed courgette flowers have become a rather trendy dish; indeed, they look wonderful on the plate, but tend to taste so delicate they may not seem worth the trouble. Geraldene Holt in *The Gourmet Garden* quotes a 1912 recipe by Mrs Yates' cook Charlotte for the surplus male flowers. The cook filled each flower with a mixture of cooked rice, minced chicken and garden herbs, tied the petals with thread and braised them in a light broth.

MARROW STUFFED WITH BLACK BEANS AND RICE

115 g / 4 oz black beans, soaked for an hour
55 g / 2 oz root ginger, peeled and chopped
1 dried red chilli, broken up
½ tsp asafoetida (optional)
55 g / 2 oz patna rice
1 medium vegetable marrow
sea salt and freshly ground black pepper
for the green herb sauce:
generous handfuls each of finely chopped parsley, chives, basil, tarragon, chervil or dill
1 tsp red wine vinegar
1 tbsp Dijon mustard
110 ml / 4 fl oz green olive oil
sea salt
little caster sugar

Boil the black beans with the ginger and chilli for about 40 minutes, until almost done. Drain and mix with the asafoetida, uncooked rice and seasoning.

Trim the stalk off the vegetable marrow, cut a canoe shape in the top so that it forms a lid, then scoop out all the pith and seeds, leaving a lining of flesh. Fill the interior of the marrow with the stuffing a replace the lid. Cover with foil and place in a baking tin, making sure the marrow lid is on top.

Cook in an oven preheated to 200°C/400°F/gas6 for 45 minutes, then let it stand and cool down a little, say for 5 – 10 minutes, before unwrapping the foil from the marrow. Slice across, serving two slices for each person. Serve with a green herb sauce made by mixing the ingredients together.

RIGHT *Marrow stuffed with black beans and rice*

MARROW WITH GREEN PEAS AND BEANS

1 medium vegetable marrow
225 g / 8 oz fresh green peas, podded
225 g / 8 oz fresh green beans, trimmed
2 tbsp sunflower oil
3 onions, chopped
85 g / 3 oz desiccated coconut
½ tsp cumin seeds
½ tsp asafoetida (optional)
1 tsp chilli powder
½ tsp turmeric
½ tsp sea salt
some coriander leaves, chopped

Peel the marrow, discard the pith and seeds, and dice the flesh into 1 cm / ½ in cubes. Bring a little salted water to the boil and plunge in the marrow, peas and beans, bring back to the boil and simmer for 3 minutes. Drain the vegetables well.

Heat the oil in a large pan and throw in the onions, followed by all the flavourings and the salt. Cook briskly for a moment, then add the vegetables, stirring thoroughly so that all are coated with the spices. Cover and simmer for 12–15 minutes.

Serve sprinkled with some chopped coriander.

STUFFED COURGETTE FLOWERS

4 small courgettes with flower attached
small can of light beer
35 g / 1 oz flour
1 tsp salt
dash of Tabasco sauce
oil for deep-frying
for the stuffing:
85 g / 3 oz cooked saffron rice
6 tbsp finely chopped parsley, chives and dill
2 tbsp fromage frais
sea salt and freshly ground black pepper

Mix all the ingredients for the stuffing together and gently fill each flower, tucking the petals underneath so as to enclose the filling.

Make a batter by mixing enough of the beer into the flour to produce a smooth runny batter. Season with salt, pepper and Tabasco.

Heat some oil for deep-frying. Dip each courgette into the batter and deep-fry for 3 minutes, until golden brown all over. Turn from time to time. Drain on paper towels. Serve with a tomato coulis.

VEGETABLE SPAGHETTI

Cucurbita pepo

Also known as spaghetti squash and spaghetti marrow, this particular vegetable appeared in our gardens only in the 1960s. It gained a high profile by the publicity it received from an enthusiast on either side of the Atlantic. In London *The Times* diarist Michael Leapman had a small allotment and would relate the vagaries of vegetable growing in his column. The golden globe of a marrow with its interior of strands that are a little sweet and fresh to the palate was one of Leapman's triumphs, and many readers wrote in to speak of their love for the vegetable. Similarly, in Los Angeles, the vegetable was made popular by Frieda Caplan of Frieda's Finest Produce Specialities, who distributed fruits and vegetables and was adventurous in extolling the virtues of rare ones. Despite these two innovators, the vegetable spaghetti is still a rare sight in the market, though gardeners grow it with enthusiasm.

There is, after all, no other vegetable quite like it. Once the strands have been disgorged from the shell and are tossed in a sauce, few people would know that they are not spaghetti until they tasted a forkful.

NUTRITION

It is very low in calories but it is an excellent source of folic acid, high in fibre and has small amounts of vitamin A, niacin and potassium.

CHOOSING AND STORING

You can eat vegetable spaghetti fresh, but it will keep, throughout the winter. If keeping for long periods, hang in a net bag from a hook, so that no part of the exterior touches a surface. However, squashes are all at their best eaten sometime in the few weeks after they have been picked. Choose ones about 20 – 25 cm / 8 – 10 in long, weighing about 1 kg / 2 – 2½ lb.

COOKING

Stick a skewer into the ends to pierce the skin and either boil the vegetable spaghetti whole or bake in an oven for about 30 minutes. You can also slice the vegetable in half and steam it with the open surface face down. Once tender, get rid of the seeds and comb the strands out of the shell with a fork on to a serving dish, pour a sauce over it and toss thoroughly.

To experience the delicate and quietly invigorating flavour, try merely adding butter and freshly ground black pepper with a little chopped raw garlic. The Sicilian method for spaghetti with *ailio crudo* serves the garlic on a little separate dish so that you can add the amount desired. Served with a well-flavoured vinaigrette or a tomato or pesto sauce, it makes a very good autumnal first course, refreshing the palate without filling the stomach.

The vegetable can be cooked whole, then cut in half (for if left unsliced it will go on cooking) and left to cool for a few hours, or until the next day. Then its strands can be turned into a salad, as follows.

AUTUMN SALAD

1 red pepper, cored, seeded and cut into strips
1 red onion, sliced thinly
1 marmande tomato, sliced in chunks
good handful of basil, chopped coarsely
1 tbsp capers
1.1 kg / 2½ lb vegetable spaghetti, cooked and cooled
for the vinaigrette:
3 tbsp olive oil
juice of ½ lemon
1 tsp Dijon mustard
pinch each of sea salt and freshly ground black pepper

Put the vegetables, capers and spaghetti into a large bowl and mix thoroughly. Mix the vinaigrette together just before serving, pour the vinaigrette over and toss the salad.

CHAYOTE

Sechium edule

Also known as christophene, chaka, pepinello and a host of other names, this is a native of Central America and the West Indies. The fruit is pear-shaped or roundish, deeply furrowed and with a single large seed. The colour varies from yellow to pale and dark green. It grows on a perennial vine with heart-shaped, angled or lobed leaves; as well as the fruits, the young shoots, leaves and large, fleshy roots are all cooked and eaten in various Mexican and West Indian dishes: the root can weigh anything up to 8 kg / 20 lb and tastes much like yam.

Buy them small and pale, so that you can eat the skin. They are bland in flavour but very juicy, so highly refreshing; the seeds can be eaten too. They keep well stored in the cool and dark. The skin on the darker ones can be peeled with a potato peeler, but if you are cooking with them leave it on.

If using them raw they are good in any of the cucumber recipes (especially the ones with spices and peanuts) or just simply sliced, blanched and served with a mustardy sweet vinaigrette. If cooking, use in combinations with tomatoes, peppers, onions and chillies, simmered together with olive oil and garlic.

The older chayotes can be stuffed like peppers with any mixtures of spicy vegetables and rice, then baked in the oven. Or they can be halved, each side scooped out and chopped with garlic, tomato and anchovy, then baked in the oven with a little Parmesan cheese sprinkled over.

In Mexico they are used in dessert dishes, in tarts, pies and cakes. *Chayote rellenos* is where the chayote shells are stuffed with spiced dried fruit.

OTHER SQUASH VARIETIES

WINTER SQUASHES

• Little Gem: dark, green and round, with pale yellow flesh

• Apple or Little Dumpling: round, with green and cream stripes; the flesh is pale yellow, and sweetish in flavour.

• Crookneck: orange to terracotta in colour with a swan neck; the flesh is orange with a slight courgette flavour.
• Golden Delicious: round and orange; the flesh is orange and slightly sweet and creamy.
• Butternut *Caryoka nuciferum*: chianti-bottle shaped; straw-coloured edible skin and dark yellow flesh.

SUMMER SQUASHES

• Acorn: pointed and dark green in colour, with yellow flesh and a mild flavour.
• Patty Pan, Custard Squash or Custard Marrow: yellow, round, flattened and fluted with white flesh, very mild.
• Scopolissi: dark green version of Patty Pan.
• Golden Nugget: round with delicate ribbing, bright dark yellow and pale orange flesh.

CUCUMBER

Cucumis sativus

Cucumber has a marvellously refreshing effect in hot climates. I remember my landlord giving me lunch at his farm when I lived on the island of Lesbos. The huge Greek meal of several courses finished with three or four large cucumbers, peeled and sliced lengthwise, to be eaten like celery sticks.

It is thought that the wild cucumber still grows in the foothills of the Himalayas and its origins lie in Asia. Certainly, some of the best recipes derive from Indian and Asian cuisine, but it was a favourite food of the Romans. The Emperor Tiberius doted upon cucumbers and had them at his table every day of the year. They were grown on trays that could be wheeled into the sun. Cucumbers were often served cooked. Apicius uses them boiled in pieces with taro in a sauce for roast crane and duck, while they are also stewed with cumin, honey, celery seed, liquamen (see right), oil and brains, then bound with egg, sprinkled with pepper and served.

Cucumbers loom large in Roman cookery, but were valued particularly for their invigorating and refreshing aspect when raw. Apicius advises serving peeled cucumbers simply dressed with liquamen, an idea which is still common in Indonesia, China and the Philippines.

The refreshing qualities of cucumber also appealed to the peoples of northern Europe. Charlemagne ordered cucumbers to be planted in the estates of his realm. We know them to have been common in England in the reign of Edward III, though they were lost soon after, when their cultivation was neglected during the long Wars of the Roses. They were reintroduced at the end of the sixteenth century, when Elinor Fettiplace gives a recipe for pickling both cucumbers and gherkins. At the same time they were less popular in France; the *Maison Rustique* believes that cucumber juice corrupts the veins and they are better kept as meat for mules. Evelyn, on the other hand, finds them cool and moist, and serves them in slices with oil mingled with lemon and orange. He also goes on to recommend a broth made from them and tells us that young cucumbers may be boiled with some white wine.

A recipe by Sarah Harrison in *The Housekeeper's Pocket Book* (1739) stews the sliced cucumbers, flours them, fries them and then serves them, with a glass of claret poured over, under roast mutton or lamb. One cannot help thinking that cooking has progressed into simpler methods, for why boil first before flouring and frying? However, cooked cucumber can be unexpectedly good, teaming up well, as one might think, with many fish dishes.

LIQUAMEN

Liquamen, the most popular condiment of Ancient Rome, was made by a sophisticated process from small fish, including anchovies, that were salted and rotted. The intense, salty, fishy essence that resulted was used much like soy sauce is today in the Orient, sprinkled on all manner of savoury dishes. We have its equivalent today in the Indonesian fish sauce called *nam pla* and various different names in the different countries that make it. The protein, minerals and vitamins it contains turn the rice with which it is flavoured into a nourishing and sustaining meal.

ABOVE Cucumber and peanut salad

NUTRITION

If you peel cucumbers they lose all their Vitamin A. They are 95 per cent water but have some vitamin C, folic acid and fibre. They are also high in potassium and sulphur.

VARIETIES

The main distinction is one of shape, since the flavour of all cucumbers is much the same. There are short, dark green 'ridge' varieties (said to be named after the ridged beds where they were grown), which have slightly pimply skins, and long, smooth-skinned kinds variously known as 'hothouse', 'English' or 'Japanese' cucumbers. You can sometimes also find 'apple' varieties – round, pale-skinned and said to be more digestible.

Gardeners in temperate countries distinguish between 'outdoor' and the 'glasshouse', 'frame' or 'indoor' varieties, chosen according to the way in which they are grown. For both outdoor and indoor cultivation, there is the option of all-female plants which are virtually seedless, and which obviate the tiresome removal of male flowers.

GHERKINS

A cousin to the cucumber, the true gherkin (*Cucumis anguria*) is native to the West Indies and parts of South and Central America. It is noted in Brazil in 1648, and was used in soups and pickles in the Caribbean. Until the nineteenth century gherkins were not cultivated, but gathered from the wild.

Gherkins are synonymous with pickles, but the distinction between true gherkins and the small, immature ridge cucumbers also used for pickling is somewhat blurred: the dill pickles that originated in Eastern Europe are often made from larger, softer 'pickling cucumbers' of the *Cucumis sativus* species.

Gherkins and small cucumbers for pickling are not always easy to find on the market. Seedsmen offer a range of gherkins, and they are as easy to cultivate and the rest of the family. A plant or two will give you over one summer masses of fruits, enough for jars of pickles and plenty to eat raw. Gherkins have comparatively thick skins, and if you decide to eat them raw, instead of pickling them, harvest them when small, no bigger than your little finger. The flavour is sensational if eaten straight after picking.

If you are bored with pickling gherkins, cook them whole, tossed in butter or olive oil and garlic. They need only a few minutes in the pan. Like boiled cucumbers, they are delicious with fish; or add them hot to a first-course salad so that the leaves will wilt.

CHOOSING AND PREPARING

Buy cucumbers when they are very fresh – firm and bright green. Beware of protective wax coatings on ridge cucumbers, which mean they have to be peeled.

I like to eat cucumbers in their season – summer. I never buy those shrink-wrapped ones which taste of nothing. A cucumber picked straight from the garden has a real and distinctive flavour. Fresh cucumbers will keep happily in the salad drawer of a refrigerator for a week. But why keep fresh food at all? It should be eaten on the day you buy it or the day after.

If you want sliced cucumber for a salad or sandwiches, the big question is whether to salt the slices beforehand. I like to extract some of the water, so tend to slice the cucumber with its peel on, sprinkle a little salt over the pieces and leave them in a colander for an hour. Then rinse the salt off under a running tap and dry the pieces before using in a salad.

Cucumber needs very little dressing. The simplest dressing would be a few drops of sesame oil, a pinch of caster sugar and a few turns of the pepper mill with a squeeze of lemon. If you want a designer effect on the salad (and don't mind losing a quarter of the vitamin A), run the prongs of a fork down the outside of the cucumber before slicing it to give a pretty striped effect. At one of the first dinner parties I gave, in my twenties, a guest looked at the cucumber and said, 'I see it died of a thousand cuts.' Unusual then maybe, but not so much now.

However, quite the best cucumber salad dishes are those from Asia. All of these make marvellous luncheon dishes with a few other salads, cooling, refreshing and stimulating. Here are two cucumber and peanut salads, an excellent combination and a favourite one in the east. The first recipe comes from Julie Sahni and has become one of our staple summer dishes. The second, from Thailand, is inspired by Vatcharin Bhumichitr.

CUCUMBER AND PEANUT SALAD

for 2

1 large cucumber
2 tbsp roasted peanuts
1 tbsp finely chopped fresh coriander
1 tbsp finely chopped fresh mint
1 tsp caster sugar
juice of ½ lemon
1 tbsp sunflower or peanut oil
1 tsp mustard seeds
1 tsp cumin seeds
½ tsp asafoetida
½ tsp turmeric
2 dried chillies, broken up
½ tsp sea salt

Cut the cucumber in half and remove the seeds unless they are very inconspicuous. Grate the flesh with the largest holes of the grater into a colander and squeeze out some of the moisture, or all that you can manage to get out.

Grind the peanuts (these can be salted or not, according to personal taste) into a powder. Place the grated cucumber into a bowl and add the peanuts, herbs, sugar and lemon juice.

Heat the oil in a pan and throw in all the rest of the ingredients until the seeds splatter (use a lid over the pan when this happens). Cook for a few seconds longer; the whole process should take no longer than a minute and a half.

Pour the contents of the pan over the cucumber and serve immediately. This is important, for if the salad is left, the cucumber will exude liquid, which will ruin the appearance and flavour of the dish.

THAI CUCUMBER AND PEANUT SALAD

1 large cucumber, grated as above
1 garlic clove, crushed
1 small red and 1 small green chilli, finely sliced
juice of 1 lemon
3 tbsp light soy sauce
3 tbsp roasted peanuts
1 large tomato, peeled and sliced

Grind 2 tablespoons of the peanuts into a powder, reserving the rest for garnish. Squeeze the grated cucumber before placing it in a bowl. Add the rest of the ingredients and toss thoroughly until well mixed. Serve at once, garnished with the reserved peanuts.

SIMPLE BOILED CUCUMBER

2 large cucumbers
30 g / 1 oz butter
300 ml / ½ pt milk (soya is excellent)
300 ml / ½ pt water
sea salt and freshly ground black pepper
2 – 3 tbsp finely chopped parsley

Trim and slice the cucumbers in half, extract the seeds and slice the flesh in 5 cm / 2 in chunks.

Melt the butter in the pan, throw in the cucumber pieces and let them sauté for a moment or two. Add the milk, water and seasoning, bring to the boil and let the pan simmer for 10 – 15 minutes. Drain, but keep the liquid for boiling new potatoes or for adding to a purée of potatoes.

Sprinkle the chopped parsley over the cucumber and serve. Or you can thicken the liquid that the cucumber has cooked in with a *beurre manié*, adding the parsley and serving the cucumber with its sauce. This simple side dish is good with fish and shellfish.

RIGHT *Marrows, courgettes with their flowers, squashes, gourds and loofahs*

GOURDS

These are the Old World cousins of the squashes. They originated in Africa, India and farther east, but are now grown all over tropical and subtropical climes. The most famous is water melon (*Citrillus vulgaris*), a native of tropical Africa, and possibly also of India. Its coral-coloured flesh is wonderfully sweet and juicy; the oily and nutritious black seeds are consumed in some places. Other gourds can be divided into bitter and mild. The bitter ones are useful in cooking, especially in Indian and Chinese dishes, giving that hint of dissonance that stimulates the palate. This bitterness is due to quinine, an antidote to malaria. It would be interesting to know whether the bitter gourds grow in malaria-infested regions, and whether the native population consumes them.

CHINESE BITTER MELON

Momordica charantia

This bitter gourd has many names, among them African cucumber and balsam pear; in India it is known as karela. Its close relative balsam apple (*M. muricata*) is slightly different. The plant is a vigorous vine which can climb up to 4 m / 12 ft. The fruits have warty skins, 'knobbly as a prehistoric crocodile', one writer thought; they change colour as they ripen from silvery white to pale and ever darker green. Fully mature they turn a brilliant orangy red. They are hugely important in both Indian and Chinese cuisine.

They are picked unripe for their bitterness. Joy Larkcom says there is a 'hint of okra and aubergine' in the flavour (which I have never detected); she further claims that the bitterness 'appears to bring out the flavour of other ingredients during cooking and is, in turn, neutralized by them.' This is certainly so.

In China the gourds are used in soup and stir-fries. They are picked while still pale green; they are cut in half, the pulp is removed, and the flesh is sliced and boiled for three minutes to remove bitterness before the final cooking. In India they are picked at a later stage. Sometimes they are sliced and salted for an hour to allow some of the bitter juices to run out; they are also often stuffed with a spiced mixture.

NUTRITION

These gourds are high in dietary fibre and have a small amount of protein and carbohydrate. They are very high in potassium, and have small amounts of calcium and magnesium and a very little iron and zinc. When ripe they are very high in carotene and have fair amounts of vitamin C and folic acid.

KARELA CHAPATTI

450 g / 1 lb bitter gourds
2 tbsp mustard oil
5 garlic cloves, chopped small
1 tsp chilli powder
½ tsp ground cumin
½ tsp coriander
1 tsp amchoor (dried mango powder)
1½ tsp sea salt
1 tsp sugar
2½ tbsp rice flour
2 tbsp yogurt

Slice the gourds with skin and seeds and boil in salted water for 5 minutes. Drain. Heat half of the oil in a pan and fry the garlic and spices for a minute. Add to the drained gourd, then mix in the salt, sugar, flour and yogurt. Let stand for 30 minutes. Heat the rest of the oil in a pan, spread the mixture over it, let the underside cook brown, then place under a hot grill to cook the top. When brown and done all through, leave to cool a little, then slice like a cake. Serve with rice and/or nan bread and other curries.

SMOOTH and ANGLED LOOFAHS

Luffa cylindrica and *L. acutangula*

The loofahs are also known as sponge gourds; both the smooth loofah (*Luffa cylindrica*) and the ribbed gourd (*L. acutangula*) are better known in the West as back scrubbers, and people often express surprise that they are eaten as vegetables. The plant is a spreading climber, which in its wild state grows rampantly over shrubs and up trees; it is thought that it was taken into cultivation only a few hundred years ago. It is very likely that the dried skeleton still hanging on the plant suggested a host of domestic uses. The smooth loofah grows to about 30 cm / 12 in long and is the sort more often used for back scrubbing, as it is easier to extract its fibrous skeleton. The angled loofah is the prettier vegetable: it is ridged and can grow up to 90 cm / 3 ft or more long.

Loofahs are eaten only when they are young; when mature they become very bitter. In China and Japan the young fruits are dried and used in winter. Use fresh loofahs as you would courgettes or cucumbers; they can be eaten unpeeled. You can cook them very simply, by slicing and frying in butter and oil for just three minutes. Or add them to a ratatouille-like mixture of peppers, tomatoes and garlic, cooked in olive oil. You can also stuff and bake them (see *Indonesian Food and Cookery* by Sri Owen).

BOTTLE GOURD

Lagenaria siceraria

This gourd was a treasured plant in the Ancient World for if a gourd is allowed to mature it develops a hard, woody shell which is still used to make bottles, spoons, bowls and musical instruments. The shape can be controlled while the plant is growing by binding the immature gourd in the way desired. Its use has been traced back to Mexican caves of 7000 BC and the Egypt of 3500 BC.

Only the young gourds are eaten; when mature they become bitter, purgative and poisonous. They come in various shades of pale green, yellow and cream. They are called '*lokhi*' in India and '*doodhi*' in Africa. Cook in the same ways as loofahs, above.

NUTRITION

The bottle gourds have only half the potassium of the bitter gourds, very little carotene and much less vitamin C. Nutritionally, in fact, they are poorer cousins.

BOTTLE GOURD *Lagenaria siceraria*

Leguminosae or *Fabaceae*

THE BEAN FAMILY

BROAD BEAN *Vicia faba*
PEA *Pisum sativum*
HARICOT BEANS *Phaseolus vulgaris*
LIMA or BUTTER BEAN *Phaseolus lunatus*
CHICK PEA *Cicer arietinum*
LENTIL *Lens culinaris*
SOYBEAN or SOYA BEAN *Glycine max*
ADJUKI BEAN *Phaseolus angularis*
MUNG BEAN *Phaseolus aureus*
ALFALFA *Medicago sativa*

It was the Romans who gave the name 'legumen' *to all edible seeds which form in pods. The great merit of these seeds to past generations was that they could be dried and stored through the winter, to be made into purées and porridges or ground into flour, but they also have great food value when consumed fresh.*

As food plants, the bean family comes a close second to the grasses. It covers the immense range of beans, peas and lentils, as well as clovers, vetches and alfalfa. A side-benefit of growing leguminous crops lies in the residues of nitrates they leave in the soil, which benefit other crops grown in rotation. By planting varieties of this family every three years, gardeners and farmers can help avoid the chemical trap.

Many of the plants not only furnish food for humans and animals, but also provide edible oils, fibres and raw materials for industry. Soya and peanuts yield oils. Fenugreek, tamarind and carob provide intense flavours. Gum arabic and gum tragacanth are used extensively in food production. Ornamental members of the family include acacia, mimosa and wisteria.

BROAD BEAN

Vicia faba

The broad bean or fava bean is one of the oldest food plants and one of the first to be domesticated. When picked in the wild state some thirty or forty thousand years ago in Central Asia, the seeds were no bigger than the little fingernail.

Legends and myths surround the bean, showing how complex and equivocal reactions of fear, respect and admiration were engendered by this simple pod. The bean was both loved and despised in Ancient Egypt. Rameses III offered to the Nile god 11,998 jars of shelled beans and, with barley, they were a constant item exported from Upper to Lower Egypt. But they were also a taboo food for some Egyptian priests, who believed that as the stems of the beans were hollow they were used by the souls of the departed as a passageway to the life after death. Pythagoras studied for at least twelve years with these priests and brought such beliefs back with him to his school at Croton on the toe of Italy. From thence, the belief that the broad bean is associated with rebirth and new life filtered into our consciousness. In antiquity beans were often offered in sacrifice in the belief that they allowed communication with the invisible world. They also became an offering in marriage ceremonies to ensure the birth of a male child.

However, aside from these mystical views of its sacred aspects, the broad bean quickly became a staple food for the masses throughout the whole of prehistory, paramount in the early civilizations and almost up to the present. It has only been superseded in the last two hundred years by the potato.

The Romans made cakes of meal from dried beans if there was a poor grain harvest; indeed, flour can be made from all pulses for cooking. The prophet Ezekiel was commanded to mix a purée of beans with the grains from which he made his bread. God obviously knew how to eke out the food supply. The word 'pulse' is from the Latin *puls*, meaning a porridge made from bean meal. Martial expresses the general love of beans in the lines, 'If pale beans bubble for you in a red earthenware pot you can oft decline the dinners of sumptuous hosts' (*Epigrams*, Book XIII). The broad bean has much going for it, since it is both highly nutritious and hardy.

In northern Europe beans can be sown in autumn for picking in mid-June and from January to April for a succession of crops. The Goodman of Paris in 1393 advises, 'do you plant them towards Christmas and in January and February and at the beginning of March; and plant them thus at divers times, so that if some be taken by the frost others be not.' If the bean overwinters in the soil, some protection is needed against mice, which burrow down and eat this store of food.

Beans have also been grown since the Stone Age as a crop for livestock. Allowed to grow large and dry before they harvested, they are thus named the 'horse bean'. I recall as a young man eating almost my first *fasoulia* in Greece and noticing how huge the beans were, I asked what variety they could be. The 'horse bean' I was told solemnly (by an elderly Hellenist) and assured it was a quite different kind from *Vicia faba* – the broad bean. I am not sure I believed him, as the bean, though white and large, was the same shape and flavour as the broad bean. But he was right. The horse, sword or Jack bean (*Canavalia ensiformis*) is grown still but mostly as a green manure or fodder crop. It is the other Old World bean, but always confused with *Vicia faba*. I wish I could get my hands on it and start growing it in my own garden. For it made the best *fasoulia* I have ever tasted.

Ever since the Middle Ages the bean played out its sacred aspect as the integral prize in the Twelfth Night cake, a custom which was common all over Europe. The cakes were made with flour, eggs, raisins and honey, flavoured with ginger and pepper. Baked in the ashes, the round, flat cakes were made as large as the household required. One portion was set aside for God, another for the Holy Virgin and three other portions for the Magi. These were all given to the poor. The rest of the cake was divided among the members of the household, and whoever had the bean in their portion became king for the day. The custom also grew up to place a pea in the cake and whoever found that became queen. Thomas Randolph (Ambassador to Elizabeth I) tells us that Lady Fleming was Queen of the Bean in 1563. One imagines that the Virgin Queen herself did not take all that kindly to these customs, but perhaps she smiled upon them – if it was just for the day.

Castelvetro, writing a few years later, recommends eating young broad beans raw at the end of a meal with a salty cheese from Crete or Sardinia, always with pepper. They are still eaten raw in Greece, but as a *mezze*; the pods are flung on the table and you break them open yourself while downing the ouzo. Castelvetro also speaks of the merits of broad bean flour for cleaning impurities in the skin.

Broad beans were classless. They were one of the staple foods of the poor, able to withstand long slow cooking and then drying, an excellent food supply for the winter. But the wealthy also enjoyed them, making rich and unctuous sauces – such as a velvety veal and chicken stock enriched with five egg yolks – to pour over the beans. Another favourite method of serving beans was to team them with bacon or ham.

FUL MEDAMES

Ful Medames is the brown broad bean which gives its name to the Egyptian dish in which it is the main ingredient. The word is thought to derive from '*mudammas*', meaning buried. A dish was cooked by being buried in hot ashes and left overnight. Ful is simply the Arabic word for fava beans. Claudia Roden writes that it is 'pre-Ottoman and pre-Islamic, claimed by the Copts and probably as old as the Pharaohs'. It is eaten all over Egypt, both in the fields and in the town, in Arab bread and garnished with tahina salad. See her *A Book of Middle Eastern Food* for recipes, and note that you can also buy them in cans.

NUTRITION

Beans are a highly nutritious food, notably high in protein and a good source of the minerals calcium and iron, vitamin A and vitamins B1 (thiamine) and B2 (riboflavin).

CHOOSING AND STORING

The fresh broad bean season is fairly short – just the summer months – so if you can, buy the pods small; the taste of the young broad bean is delicate and very fine. If the pods themselves are small enough, no longer than 10 cm / 4 in, you can cook them whole. Even the tender leaves are edible, cooked like spinach or any greens. Gardeners who pinch out the tops of the plants to deter blackfly and stimulate bean development might try this tasty alternative to composting the leaves.

Try and cook broad beans as soon as they are picked, for the pods will grow limp within two days and the sugars in the beans turn to carbohydrate, changing the flavour and texture. If you have a garden, however small, I would recommend growing your own. Freshness is all.

Having said that, the broad bean is the vegetable that suffers the least harm when frozen, provided the freezing is done when the crop is freshly gathered. Though still nowhere near having as much flavour as fresh, frozen beans can be used in all the recipes here. As a passionate consumer of food in season, I myself would never eat frozen beans outside the summer – which I must admit rather takes the point away from freezing them at all.

VARIETIES

Seedsmen's catalogues distinguish between green and white, round and kidney-shaped beans, as well as those with longer and shorter pods. Gardeners should choose the varieties that suit them – in some areas, for example, autumn sowing is a possibility; there is little difference in the final flavour whatever the chosen variety.

The oldest variety and still a favourite is Green Windsor, cultivated first at Windsor by Dutch gardeners. (Up to the end of the last century there still existed a garden near Eton called the Dutchman's

garden.) Sutton Dwarf only grows up to 45 cm / 18 in tall and therefore needs no staking. Both of these have shortish pods. Other varieties have long pods:
AQUADULCE is the favourite bean for autumn sowing.
RELON is the biggest and longest; each pod can grow to 50 cm / 20 in.
HYLON is a newer variety which is hardy and long.
EXPRESS is an early, fast-maturing bean which is also a heavy cropper.

PREPARING AND COOKING

If you are fortunate enough to be in possession of young beans, cook them whole. Simply top and tail, rinse under a tap and boil for 4–5 minutes. The pods are unexpectedly filling, so you will find a pound in weight (450 g) will happily satisfy six to eight people as a side dish. Another way with the young pods is to make a soup. Simply cook the pods, without the beans, in water or a light vegetable stock for 4 minutes, liquidize and then push the thin purée through a sieve to eliminate the fibres. This soup will, of course, be richer and thicker if you include the beans as well as the pods. Finish the soup with a knob of butter or a swirl of sour cream.

When the beans become larger and more mature, the question is whether you have to peel the outer skin on the bean itself? Dorothy Hartley puts it precisely, quoting country people: 'when the placenta (i.e. the little hook that fastens the bean to its pod) comes loose easily and leaves a white scar, the bean is still young enough for ordinary cooking, but when the scar becomes black, then the country people call them "blackspotted beans" and reckon them indigestible food, though good enough for a hungry plough-boy – and indeed, some old country folks do not "reckon you get the full flavour of a broad bean till the black spot is come and the grey skin is pretty tough".'

Most of the broad beans sold in the shops need no peeling, except at the end of the season. This seems a fiddly job to some, but the flavour of the bean is so special that it is worth it. The job is done with less hassle if the beans are boiled 2 minutes, left to cool, then peeled, rather than attempting to peel the new bean. The peeled beans are then finished by whichever particular cooking method is chosen.

Another way to remove the outer skins is to cook them unpeeled, then mash the beans and push through a strainer. This is an age-old method of making the bean meal or flour for bean cakes, breads and croquettes. Andrew Boorde (1490–1549), the physician and traveller who was also Bishop of Chichester, comments 'beene butter is use moche in Lent in dyvers countries – it is good for plowmen to put in their paunches.' Hartley adds that this was 'dried winter beans boiled to a mush in mutton broth and used as a thick spread upon coarse oatcake.' A rather more refined version is to mix the ground beans with butter, yogurt or sour cream, to flavour the mixture with garlic, parsley or savory and to use it as a spread. You can also add a beaten egg and fry small cakes of the mixture as appetizers.

A Sardinian recipe – *faiscedda* – cooks the bean purée (flavoured with nutmeg and cinnamon, stiffened a little with breadcrumbs and held together with several eggs) in a pan as one large cake. It is fried on both sides, and when cooked through, sliced in wedges.

The classic French flavouring for broad beans is the herb savory. If you use it, be circumspect as the herb can easily overpower the bean. Not, one would think, an easy thing to do. My own preference is to serve broad beans with parsley sauce. Nothing is simpler or more mundane, you might think, but the fusion of flavours is enormously satisfying. Ensure the white sauce itself is rich in butter and very low in the thickening power of the flour and that the parsley is chopped very small, so that it flecks the sauce like green speckly threads.

RIGHT Broad bean, lentil and redcurrant salad (overleaf) with runner beans in green peppercorn sauce *(see page 168)*

BROAD BEAN, LENTIL AND REDCURRANT SALAD

for 10

This recipe makes a wonderful summer salad as part of a luncheon or party dish.

1.35 kg / 3 lb young broad beans
225 g / 8 oz brown lentils
bunch of chives or spring onions
bunch of basil, chopped
1 garlic clove, crushed
5 tbsp olive oil
1 tbsp red wine vinegar
1 tsp Dijon mustard
pinch each of sea salt and freshly ground black pepper
2 punnets redcurrants

Pod the beans and cook them in boiling salted water for 4 minutes. Drain and reserve.

Boil the lentils in a large pan of boiling water for 10–12 minutes. Drain and reserve.

In a large serving bowl, mix beans and lentils while still warm. Chop chives or spring onions finely, and add to the bowl with the basil.

Make a vinaigrette by mixing the crushed garlic, oil, vinegar and mustard with salt and pepper to taste. Pour over and toss well.

Lastly, take the stalks from the redcurrants and mix those in too, carefully leaving a few to show jewel-like on the top.

BEAN TARTLETS WITH OEUFS MOLLETS

450 g / 1 lb broad beans
bunch of parsley, finely chopped
bunch of chives, finely chopped
4 prebaked shortcrust tartlet cases
4 eggs
1 tbsp green peppercorns in brine
150 ml / ¼ pt sour cream
sea salt and freshly ground black pepper

Pod the beans and boil them for 4 minutes.

Liquidize, using a little of the water they have cooked in, to make a smooth thick purée. Stir in the chopped parsley, chives and seasoning.

Divide the bean purée among the tartlet cases.

Boil the eggs for 4 minutes. Immediately plunge under a cold tap and leave to cool. Tap each egg gently with the back of a spoon until the shell is finely cracked, then peel under running water. Dry the eggs (which should be thoroughly *mollet* or soft-boiled) and place one on top of each tartlet.

Finally, mix the drained green peppercorns with the sour cream and spoon a generous amount of this mixture over each egg. Serve as a first course with a little red lettuce garnish.

BEANS AND COURGETTES IN WINE

675 g / 1½ lb broad beans
4–5 courgettes
30 g / 1 oz butter
1 tbsp oil
1 onion, thinly sliced
175 ml / 6 fl oz dry white wine
salt and freshly ground black pepper
2–3 tbsp chopped parsley

Pod the broad beans and trim and slice the courgettes diagonally.

Melt the butter in a pan and add the oil, throw in the onion, then add the beans and courgettes. Sauté for a moment or two, turning the pieces in the fat.

Add the wine and seasoning. Place a lid on the pan and let it simmer for 5 minutes.

Pour into a serving dish and sprinkle with the chopped parsley.

PEA

Pisum sativum

The pea was yet another vegetable that was domesticated very early in prehistory – possibly in India, for the name is Sanscrit. The original species would have been *Pisum arvense*, thought to be the original pea which the others all stem from. We still grow and eat *P. arvense*, but know it in the kitchen as the split pea (it is referred to also as the field pea and grown for cattle fodder). Both this and *Pisum sativum* were eaten in Ancient Egypt – seed has been discovered in the tombs at Thebes.

However, though peas are mentioned several times by Roman writers like Columella and Pliny, they do not seem to have exerted much attraction for them. They are lumped in with the rest of the legumes and one feels that peas must also have been used for porridge. As early as the fifth century BC, Greek plays mention 'pease porridge, beautiful and brown'. So it was certainly *Pisum arvense* that was grown extensively and became a staple part of the diet for the common people, cooked for long hours in soups and stews, dried in winter and used to make pease pudding.

The green pea or garden pea which excites us took a much longer time to provoke a writer. A Frenchman at the court of Louis XIV in 1695 complains, 'It is frightful to see persons sensual enough to purchase green peas at the price of 50 crowns per litron.' The French court at this time was besotted by green peas. Madame de Maintenon wrote in a letter dated 10 May 1696, 'this subject of peas continues to absorb all others. The anxiety to eat them, the pleasure of having eaten them and the desire to eat them again, are the three great matters which have been discussed by our princes for four days past. Some ladies, even after having supped at the Royal table and well supped too, returning to their own homes, at the risk of suffering from indigestion will again eat peas before going to bed. It is both a fashion and a madness.'

More than a hundred years or so later the madness had not yet faded. Exorbitant prices were paid for the first green peas of the season, Brillat-Savarin notes that the first plate of green peas cost about 800 francs – equivalent to £32 at today's rates.

Nor were peas valued for their simplicity; they appear to have been cooked in quite a complex manner. Boiled first for 20 minutes, they were drained, then butter and flour were added with some pepper, salt and sugar; the mixture was simmered for five more minutes, then the dish finished with cream and several egg yolks.

The *Cuisinier Gascon* (Amsterdam 1740) on the other hand gives a recipe for a *Macedoine à la Paysanne* which sounds very contemporary, except for the amount of butter. It is peas, broad beans and green beans (both diced to the size of a pea), and carrot slices, all cooked in butter together. That gourmet Grimod de la Reynière called peas 'the gayest song of the month of May'.

All this enthusiasm for the pea stimulated new types and hybrids (Thomas Jefferson grew thirty different varieties in his Virginian garden at Monticello), many of which were lost a hundred years later.

MANGE-TOUT OR SUGAR PEA

These peas produce thin, flat pods that are ready when the pea shape just shows through the pod wall. The whole pod is eaten before the seeds swell – hence the name mange-tout or 'eat-all'. Also known as snow pea or Chinese pea, this type is associated with Chinese and Japanese cooking, but originated in the Mediterranean region. Gerard mentions the sugar pea in 1597, and it was known in France sixty years before the advent of the green pea that so delighted Louis XIV. Some years later a writer complains that it has disappeared, but it periodically reappears in the records and is nowadays firmly established in greengrocers' displays and seedmen's lists.

PETIT POIS

Petits pois are the fruit of dwarf varieties specially bred to produce tiny but highly sweet peas. Rather than buy them tinned or frozen, which is how petits pois are marketed nowadays, you can always grow them in your own garden. Boil in the pods and shell before serving.

ASPARAGUS PEA

Lotus tetragonolobus

This is actually a type of vetch rather than a pea proper. The fruits are small ridged wings, which have to be gathered when they are tiny – no bigger than 2.5 cm / 1 in – otherwise they are fibrous and inedible. Even when small they are not much liked, and for once I agree with the critics. The flavour is sweet and distinctly pea-like, but they have to be picked at exactly the right moment.

VARIETIES

Apart from the asparagus pea, the different types are all varieties of *Pisum sativum*, bred for slightly different characteristics: mange-tout, intended for picking before the seeds swell; petits pois, where the mature peas remain tiny, and a range of conventional types. Here gardeners can choose between a wealth of round and wrinkled or marrowfat varieties according to when they want to sow and harvest their crops, selecting ones with a dwarf or taller-growing habit to suit their space. Mange-tout varieties include medium-height Edula and Oregon Sugar Pod and the taller Sugar Snap. For petits pois, grow dwarf Waverex or slightly taller Gullivert.

NUTRITION

Green peas and their pods are high in protein, carbohydrates and vitamins B and E as well as calcium, phosphorus and potassium.

CHOOSING

Fresh garden peas have a very limited season, alas, so make the most of them. It is tragic that every year one hears of children who on seeing the fresh pods in the market enquire what on earth they are, so convinced have they been that the fresh pea comes frozen in packets.

Look for pods which are a clear fresh green, not dull, and which have a bouncy shape, never drooping. Cook them on the day that you buy them. When the

RIGHT *Peas, runner and french beans*

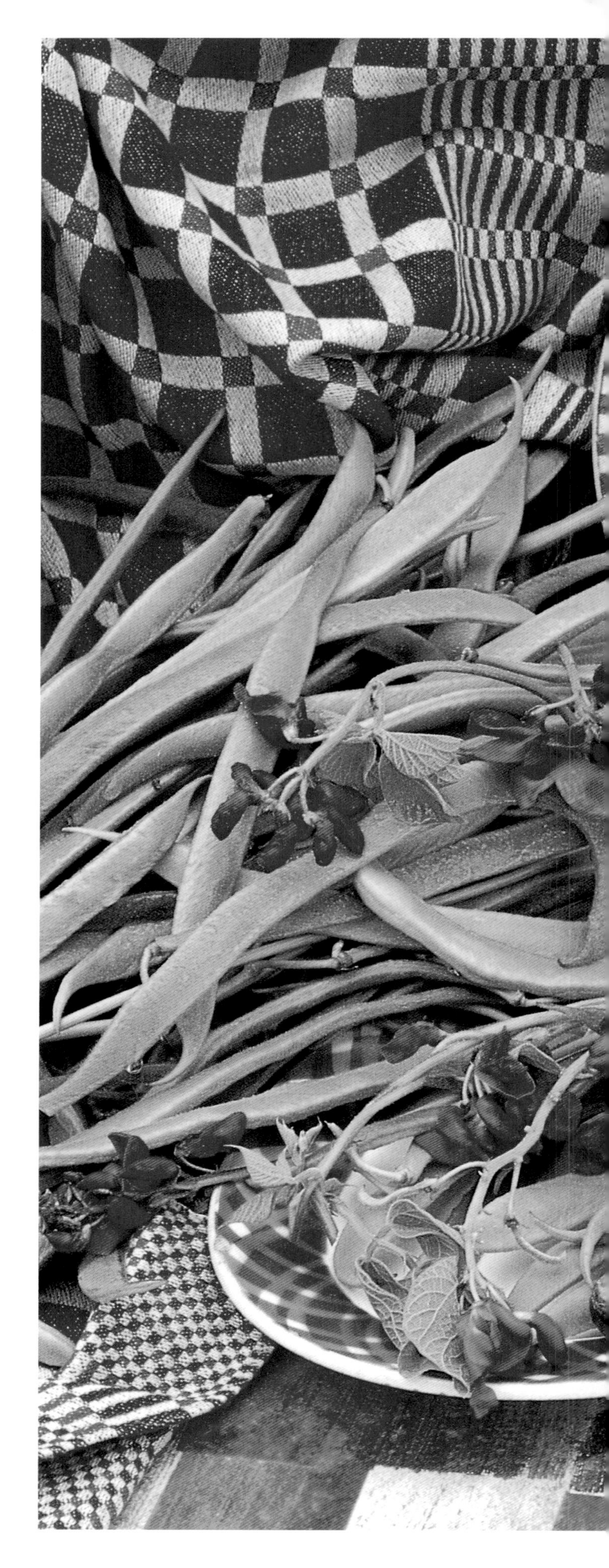

ABOVE Fresh green pea soup

fresh green colour has turned drab the peas are older; perfectly edible and delicious, but the sugar has turned to carbohydrate and they will have that slightly mealy consistency of the pulse.

If you want to make your own dried peas, leave them on the plant until the pods become husk-like, then pick and pod the peas, spreading them out to dry on trays left in a cool, dark place.

COOKING

If it is possible, pod garden peas immediately before cooking. Podding is a fulfilling task which most people enjoy, so share it among friends. Fresh peas need to be simmered in boiling water with a little salt and sugar for about 2–3 minutes. Drain and add a pat of butter. Some cooks like to add a few pods to the water and a sprig of mint. It is a matter of taste: if you find these improve the flavour, use them.

Mange-tout peas are cooked very simply, either by steaming for 3 minutes or boiling for 1, or by stir-frying for about 2 minutes. They can also be eaten raw and are extraordinarily sweet and refreshing. You can also make an excellent soup out of them if you are lucky enough to have a surplus harvest. The soup is best eaten iced. You would need a good pound in weight to make enough for 2–3 people. Just trim the pods and boil in water or stock for 3–4 minutes, then cool and liquidize. Put through a strainer in case there are fibres, season, then add 300 ml / ½ pt buttermilk or smetana, mix in thoroughly and refrigerate for 2 hours.

One of the classic methods of cooking garden peas, used by both Soyer and Dumas, is to cook them on a bed of lettuce, so that they are steamed. They also add white wine and copious amounts of butter and cream. It is a good method (see below) but, on the whole, peas need little added to them; they are magnificent in themselves. Their flavour is most trenchant in the following soup.

FRESH GREEN PEA SOUP

For six

900 g / 2 lb peas in their pods
1.5 litres / 2½ pt vegetable stock
1 sprig of mint
½ tsp sea salt
½ tsp sugar
300 ml / ½ pt buttermilk, smetana or single cream
mint leaves for garnish

Pod the peas and cook the pods in half the stock for 5 minutes. Leave to cool, then liquidize and pass through a strainer. Reserve the liquor and throw away the pod debris.

Cook the peas in the rest of the stock with the mint, salt and sugar, for 3 minutes. Leave to cool, discard the sprig of mint and liquidize. Combine this mixture with the pod liquor.

Taste for seasoning, then mix in the buttermilk, smetana or single cream. Refrigerate if serving the soup iced and decorate with the mint leaves. If serving the soup hot, reheat with care and take off the flame before it boils.

POIS ST GERMAIN

2 cos lettuces
2 - 3 tiny carrots, sliced
1 onion, sliced
2 - 3 spring onions with their green, chopped
2 glasses of dry white wine
900 g / 2 lb peas, podded
pinch of sea salt
pinch of sugar
30 g / 1 oz butter
chopped parsley, for garnish

In a large saucepan, lay the whole lettuce leaves so that they entirely cover the bottom, then add the carrots, onions and wine, and finally the peas, seasoning and butter. Place the saucepan over a low heat, fit a lid firmly on the top and forget about it for 20 minutes, when all should be cooked and tender.

Pour into soup plates, garnish with parsley and enjoy for lunch with good crusty bread. Quite my favourite way of cooking peas.

Note: An excellent purée of peas can be made by omitting the carrot and liquidizing the above, adding at the last moment 2 tablespoons of thick cream and plenty of freshly ground pepper.

SOUFFLÉ DE POIS VERTS

900 g / 2 lb peas, shelled
300 ml / ½ pt vegetable stock
55 g / 2 oz butter
30 g / 1 oz flour
300 ml / ½ pt milk
15 g / ½ oz Gruyère cheese, grated
15 g / ½ oz Parmesan cheese, grated
4 eggs, separated
sea salt and freshly ground black pepper
1 tbsp double cream

Preheat the oven to 220°C/425°F/gas7 and butter a large soufflé dish.

Boil the peas in the stock, and after 2 minutes take out half of them with a slotted spoon and reserve. Continue to boil the rest of the peas for another 3 minutes, then season, liquidize and reserve as the sauce.

Melt the butter in a pan, add the flour and make a roux. Season, then add the milk and cheeses until you have a thick, smooth sauce.

Take away from the heat and beat in the 4 egg yolks, then add the green peas.

Beat the egg whites until they are stiff and fold them in to the mixture. Pour the mixture into the soufflé dish. Bake in the centre of the oven for 20 minutes, until golden brown and risen.

Serve at once with the pea sauce, reheated, adding the double cream at the last moment.

HARICOT BEANS

Phaseolus vulgaris

This is a huge genus, packed with familiar names, but basically all are the same bean. Whether we call it kidney, French or runner, whether it is oval, flat or round, whether pink, purple, green or black, whether the pods are smooth, short, straight or curved, whether the whole pods are eaten fresh or the beans alone eaten dried, the haricot bean has proved to be astonishing in its variety.

Columbus was possibly the first European to have encountered the bean in Cuba, but it was rediscovered in the next few decades in Florida, in South America and as far north as the mouth of the St Lawrence river. This shows us how ancient its use as food was (the bean has been discovered in Peru dating from 8000 BC and in North America from 5000 BC), for the bean had time to adapt itself to contrasting climes and habitats. The first settlers were hugely impressed by this food; 'their ordinairie food is of pulse, whereof they have great store, differing in colour and taste from ours', says John Verrazano in a letter written in 1524. In 1605 Champlain observed that when the Indians planted corn they put in each mound three or four Brazilian beans of different colours: the beans used the growing corn for support. Captain John Smith in 1614 noticed that the Indians kept bags of beans stored away in a pit.

The European herbalists named them all kidney beans, believing they strengthened the kidneys. They based such a conclusion, of course, on the Doctrine of Signatures, where a kidney-shaped bean must do good to a kidney-shaped organ. The word 'haricot' is a corruption of the Aztec word '*ayecott*'. A *haricot de*

BELOW *Yellow runner beans, french beans and peas*

mouton, a dish of mutton stew with turnips, onions and potatoes, is now served also with haricot beans as if to fulfil the name. But this 'haricot' comes from the French *harigoter*, meaning to cut up small, and one can also have a 'haricot' of beef and game, all cooked and served without beans.

Once the new beans had reached Europe it did not take long before they were grown. They were eaten green to begin with and were most popular in Tuscany. The bean was known in France and England as the Roman bean while Tuscans became '*manga fagioli*', 'bean eaters'. Runner beans were grown as decoration, clambering up poles made into arbours around places to picnic or banquet. Though much liked when eaten by the rich for their flavour, they were also frowned upon as the cause of flatulence and for this reason, perhaps, horticulturalists did not grow many of the different varieties that were available.

It was not until the late nineteenth century that a vast number of different beans was grown commercially. Early beans were forced under glass; Cobbett notes that plants would come into flower ten days after germination. The types grown for drying did not catch on in Britain, but were grown extensively in France, Spain and Italy. This has given rise to the idea that such beans need a warmer clime in order to flourish. This is not true as long as they are sown after the last frost. I grow both haricot and flageolet in my own garden in Suffolk very successfully and, as we have seen, some types of haricot grow happily as far north as the area of the Great Lakes in Canada.

NUTRITION

All beans are rich in protein, carbohydrate, minerals and vitamins. Dried beans have greater nutritional qualities than green beans, but less vitamin C. While all the beans have differing amounts of goodness they are all rich in lysine, which is the essential amino acid that is lacking in grains.

All the dried beans are also high in the indigestible starches, stachyose and raffinose, which sound like some appalling music-hall act. These produce the wind in the gut that causes some people to avoid all dried beans completely. These two starches do not pass through the walls of the small intestine to become converted into blood sugar; instead, the intestinal bacteria break them down into carbon dioxide and hydrogen – an explosive mixture.

VARIETIES

Green beans

French beans and runner beans: No one, to my knowledge, has discovered why the former bean, as its name implies, is popular in France and why the latter is never grown. I believe the runner bean has more flavour, which makes French lack of enthusiasm for it even more puzzling. Because it is a climber, the runner bean is never grown commercially, for it is too difficult to harvest. This might explain its immense popularity in British cottage gardens. Runner beans are far from hardy: plant them late, well after all risk of frost. They should be eaten young, before the beans inside have grown bigger than a tiny finger nail. They can then be cooked whole – simply topped and tailed, as with the French beans. But if left to grow larger, they can be sliced. The British tend to slice them thin, on the diagonal, a habit that does nothing for flavour or appearance. Better, by far, if the bean has reached a size where the pod is fibrous, to leave it to dry out. For the dried beans can be used that season. They take only 10 – 12 minutes to cook and are full of flavour, apart from being prettily marked and looking attractive on the plate. Serve them with butter and a little chopped onion, or with sesame oil, lemon and garlic. Runner beans also go well with a garlicky tomato coulis, with almond butter, or with a cream and green peppercorn sauce (see page 168).

French beans are variously known as snap bean, bobby bean and wax bean. They are always eaten young, as whole pods with immature seeds. Steam or boil them briefly, or serve them with any of the sauces above.

French beans: These have now been bred as compact bushes 30 – 45 cm / 12 – 18 in high. They come in green, yellow and purple pods, often stringless. The most popular varieties are Prince, Tendergreen, Loch Ness and Remus. Two climbing French beans are Blue Lake and Garrafal Oro, but there are over 40 listed in a seed catalogue against the large number of eighty-four dwarf varieties. Two yellow varieties are Mont d'Or and Kinghorn, while two purple beans are Royal Burgundy and a purple climber – Purple-podded Climbing.

RUNNER BEANS *Phaseolus coccineus*: The climbing beans will all grow 2.5 – 3 cm / 8 – 10 ft high and need stick or trellis supports. Varieties are Achievement, Red Knight, Streamline, Scarlet Emperor and Kelvedon Marvel. Dwarf runner beans are Hammond's Dwarf Scarlet and Hammond's Dwarf White.

BEANS FOR DRYING

Many of these beans straddle the divide between the 'green' and 'dried' categories, depending on when you harvest them. In most cases the fresh pods can be eaten whole (at least by the gardeners who grow them), and the shelled mature beans can be eaten fresh or dried. Some of the following varieties appear in seedsmen's catalogues; most can be found on food store shelves either canned or dried in packets. In late summer and autumn the season's crop can sometimes be found, semi-dried, in ethnic shops or Continental markets.

FLAGEOLET: Called thus because the French decided it resembled a flute. This bean stands supreme in its reputation as the essential gourmet bean. Either pale green or white, it can be bought in French markets still in its dried pod around September/October. These beans are so fresh they do not need soaking and will cook within 20 – 30 minutes. Once they are dried out and have been stored for a year or more, soak for an hour and give them another hour's boiling.

NAVY BEAN: A small white haricot, also named Pearl Haricot, which has been specially bred for the baked bean industry. It can be used for home cooking though, on the whole, I would tend to use the larger haricot varieties. Nevertheless, for an authentic home-made baked beans dish, use this one.

RED KIDNEY BEAN: A popular bean for party food, as its colour is dramatic. In the sixties it caused some controversy when foolish people served salads where the beans had been soaked and not cooked. One's imagination boggles that the guests actually ate them. But some did and became ill with a toxin (haemagglutinin) present in the raw bean which can cause gastroenteritis. The beans must, after soaking, be boiled vigorously for 10 minutes, then the water is thrown away and the rest of the cooking begins.

BLACK BEAN or TURTLE BEAN: This is not the bean of Chinese black bean sauce, which resembles the adjuki bean, but is much used in Caribbean cookery and in Mexico, as a soup and a sauce.

BORLOTTI BEAN: The most popular Italian bean, this can be a pale creamy pink to beige and brown speckled like a bird's egg. Pinto beans are smaller but very similar and are favoured in Spain and the US where they are grown.

BIANCO DE SPAGNA: Another firm Italian favourite, this is a large, white bean with a similar flavour to the borlotti. Canellini is another bean which is cooked with the bollito misto. This originated in Argentina but has now become one of the most familiar beans in Italian cuisine.

COOKING DRIED BEANS

If the bean is picked when just dried from the plant, it needs no soaking and only a mere 15 minutes' cooking by boiling in water with *no* added salt. However, few of us can harvest our beans in this manner; the nearest we get to it is when buying white haricot or pale green flageolet from French markets in September and October. But if you buy your dried beans in France and Spain through the winter months, you will still find they are that year's crop and need at the most 30 minutes' cooking.

If buying your dried beans in the shops, then always soak them. There is no need to soak them overnight. I favour blanching the beans with boiling water and leaving them for an hour, before pouring the water off and starting the cooking. Tom Stobart explains: 'The objection to soaking is that it starts enzyme changes which would lead to germination; in warm weather, it may also start fermentation by micro-organisms. This is obviated by putting the beans into boiling water, blanching them for 2 minutes – which upsets the enzymes, kills the beans and most of the organisms – and then soaking them for 1 or 2 hours.'

On the delicate subject of the indigestibility of beans, much has been written. I used firmly to believe

that changing the soaking water three times before cooking helped, or that cooking with various antiflatulent herbs, an old German remedy, would remove the hazard. But nothing actually really does the trick. If you love beans as I do, you merely have to resign yourself to coping with a certain amount of wind and hope that spouses and partners will toot along in a companionable way.

Once you know why the wind exists it helps, I believe, that necessary resignation to the inevitable. The cause is that some of the sugars in the bean's starch (the oligosaccharides) are left untouched by the enzymes in the digestive system, so they leave the upper intestine unchanged. Once in the lower intestine the bacteria go into action and break down the sugars into various gases. As Harold McGee points out, 'the oligosaccharides are especially common in seeds because they are one form in which sugars can be stored for future use.' They also tend to accumulate in the seeds' final stages of growth, which is why green beans give no trouble. Both cooking in water and sprouting the seeds will decrease the amount of oligosaccharides.

Beans need to be boiled for anything from 30 to 90 minutes; it all depends on the bean (flageolets need less) and how old they are. But the most common time is around an hour, after which they will be soft enough without collapsing into an unsightly mush. Always add salt at the end of the cooking or the skins will toughen. For most recipes I would start the soaked beans in a little olive oil, a sprig of bay and rosemary and perhaps a sliced onion and a clove of garlic. Remember at the end of the cooking to remove the herbs. Beans and olive oil are natural companions and most bean dishes, whether served hot or cold, will need a little good-quality oil poured over them.

All types of fresh and dried bean make superior soups. Here are my two favourites, which never fail to please. The second is my recipe to celebrate spring, a particularly rich, vibrant, green soup with the strong lemony flavour of sorrel and the slight pepperiness of the watercress as foils to the flageolets. The first recipe is my version of the classic bean soup of Tuscany.

TUSCAN BEAN SOUP

170 g / 6 oz large haricot beans, soaked
3 – 4 tbsp olive oil
1 head of garlic, peeled
2 tbsp fresh oregano, chopped
2 medium potatoes, peeled and diced
2.55 litres / 4½ pt water or vegetable stock
handful of parsley, chopped fine
sea salt and freshly ground black pepper

Drain the beans. Heat the olive oil in a large saucepan and throw in the whole garlic cloves (reserving two), sauté for half a minute before adding the beans, the oregano and the potato. Continue to cook, stirring, for another half a minute, then add the water or stock, place a lid on the pan and simmer for an hour when the beans will be tender.

Season, then ladle half of the soup into a liquidizer and blend. Return this purée to the saucepan. If the soup is too thick and chunky, add more water or stock. At this stage, vegetable stock is best as it will not dilute the flavour.

Chop the two reserved garlic cloves, mix with the finely chopped parsley and add after reheating the soup, just before serving.

BAKED BEANS

170 g / 6 oz navy (small white haricot) beans, soaked
4 tbsp olive oil
1 whole onion, sliced
3 garlic cloves, sliced
2 large tins (400 g / 14 oz) of chopped tomatoes
3 tbsp muscovado sugar
150 ml / ¼ pt red wine
2 tbsp tomato paste
1.75 litres / 3 pt water
1 tbsp dried mustard powder, mixed to a paste with a little water
sea salt and freshly ground black pepper

Preheat the oven to 160°C/325°F/gas3.

Drain the beans and heat the olive oil in a large casserole dish, throw in the onion and garlic and sauté for half a minute, then add the beans and cook for another minute, stirring all the time.

Add the tomatoes, wine, tomato purée and water and bring to simmering point. Place a lid on the casserole dish and bake in the oven for 3½–4 hours.

Take the beans from the oven, stir in the mustard paste and check the seasoning.

SORREL, FLAGEOLET AND WATERCRESS SOUP

115 g / 4 oz flageolet beans, soaked for an hour
2 tbsp olive oil
1 onion, thinly sliced
1.75 litres / 3 pt vegetable stock
55 g / 2 oz butter
2 or 3 courgettes, thinly sliced
115 g / 4 oz sorrel leaves
1 bunch of watercress
300 ml / ½ pt dry white wine
½ tsp sea salt
300 ml / ½ pt buttermilk or smetana

Drain the beans, heat one tablespoon of the olive oil in a large pan, throw in the sliced onion and the beans, let them sweat for a few seconds, then add the vegetable stock. Bring to the boil and simmer for half an hour or until the beans are cooked. Let them cool, then liquidize half of the beans, return this purée to the pan and reserve.

Meanwhile, in half the butter and the rest of the olive oil, sauté the courgettes for 10 minutes, then add these to the soup.

With the rest of the butter cook the sorrel leaves (take out the central spine if the leaves are big) for a few minutes until they turn into a dark khaki purée, then stir this into the soup.

Next chop the washed watercress with a generous amount of stalk and place in a blender jar with 150 ml / ¼ pt of the dry white wine. Liquidize this mixture to a smoothish purée.

Add to the soup with the rest of the wine. Add the sea salt and taste. Heat gently again and just before it reaches simmering point, add the buttermilk or smetana. Stir and serve.

THREE-BEAN SALAD

for 6 to 10

for the flageolets:
170 g / 6 oz flageolet beans, soaked
3 tbsp olive oil
3 garlic cloves, peeled
1.75 litres / 3 pt vegetable stock
handful of parsley
bunch of watercress
3 tbsp walnut or hazelnut oil
juice and zest of 1 lemon
sea salt and ground black pepper
2 leeks, very finely sliced
for the kidney beans:
170 g / 6 oz red kidney beans, soaked and cooked
4 tbsp olive oil
3 red peppers, cored, seeded and sliced
1 red chilli
3 garlic cloves
for the black beans:
170 g / 6 oz black beans, soaked
85 g / 3 oz root ginger, peeled and chopped
3 tbsp olive oil
3 garlic cloves, peeled and sliced
bunch of spring onions, chopped
2 tbsp sesame oil
juice and zest from 1 lemon

For the flageolets: Drain the beans, heat the olive oil in a saucepan and throw in the garlic and the beans, sauté for half a minute then add the vegetable stock, bring to a simmer and cook for 30–45 minutes, or until the beans are tender.

Pour off about 250 ml / 8 fl oz of the liquid into a blender jar, add the parsley and the watercress, the walnut or hazelnut oil, the juice and zest of the lemon and seasoning. Blend to a green sauce.

Drain the flageolets (save any liquid for soup) and add the green sauce and the finely sliced raw leeks.

Stir and leave to absorb the flavour for 1 hour.

For the kidney beans: Drain the cooked beans and reserve. Heat the olive oil and add the red peppers, chilli and garlic, cook over a low heat for 10–15 minutes, then liquidize everything into a sauce.

Add some seasoning to taste, then pour the red pepper sauce over the kidney beans.

Leave for at least an hour to allow the beans to absorb the flavour of the sauce.

For the black beans: Drain the beans, heat the olive oil and throw in the ginger and garlic. Cook for half a minute before adding the beans, cook for another minute, stirring the mixture, then add 1.75 litres / 3 pt of water. Simmer the beans for an hour.

Drain, then add the chopped spring onions, sesame oil, lemon zest and juice and seasoning.

Leave for an hour to absorb the flavours.

Serve the three-bean salad in a pattern of circles or triangles on a large round serving dish. Unfortunately, once the guests start serving themselves the pattern tends to blur, but it looks attractive at the beginning and the flavours contrast beautifully.

RUNNER BEANS IN GREEN PEPPERCORN SAUCE

45 g / 1 lb small runner beans
(mixture of green and yellow if possible)
150 ml / ¼ pt crème fraîche, fromage frais, smetana or sour cream
1 tbsp green peppercorns in brine
handful of basil, chopped

Top and tail the beans, or if large, slice them in two and make sure the sides are without fibre. Steam for 5 minutes, or boil for three. Drain them well.

Meanwhile, mix the cream, fromage frais or smetana with the peppercorns and stir in the chopped basil.

Leave for 10 minutes or so, then serve the beans with the sauce spooned over them, still slightly warm.

LIMA OR BUTTER BEAN

Phaseolus lunatus

One of the largest beans we have, this originated in South America but is now grown all over the world. It is sometimes called Madagascar bean. Lima beans prefer a moist, warm climate, so cannot be grown in temperate countries. The bean was first confused by early settlers with another type of haricot, for lima beans come in all colours and patterns. The prettier forms marbled in red and pink are often slightly toxic (hydrocyanic acid), which is why we see only the flat round white ones that are grown commercially. In the USA lima beans are eaten fresh, straight from the pod like broad beans; they are also sold frozen and canned. In the USA they prefer a lima bean that is smaller than the butter beans sold in Britain.

When dried, lima beans need an hour's soaking and will take up to 45 minutes' simmering until they are tender. They have a very pronounced flavour, quite different from the range of haricot beans, and make excellent dishes. They have one drawback: when cooked, the outside skin often peels away, making them look rather unsightly. One way of avoiding this is to turn the cooked beans into a purée which can be eaten with warm pitta or garlic bread rather like hummus, as in the following recipe.

BUTTER BEAN PURÉE

140 g / 5 oz butter beans, soaked
3 garlic cloves, crushed
juice and zest of 1 lemon
75 ml / 5 tbsp olive oil
handful of chopped mint
sea salt and freshly ground black pepper

Boil the beans in plenty of water for 45 minutes, or until tender. Place in a blender and add the lemon juice and zest, oil and seasoning. Liquidize to a smooth creamy purée, stir in the mint and serve.

RIGHT *Baked beans (page 166) and three-bean salad (previous page)*

CHICK PEA

Cicer arietinum

Nicknamed 'owl's head' in Egypt, this staple food in the ancient world is known to have been cultivated as early as 7000 BC. Both chick peas and lupin seeds were sold hot in the streets; Horace describes them as part of an 'economical diet of onions, pulses and pancakes'. The Latin name, *cicero*, like the broad bean (*fabius*) and the lentil (*lentulus*) was used by several distinguished Roman families, which shows how popular it was. There were then many varieties, differing in size and colour. Albertus Magnus in the thirteenth century mentions red, white and black chick peas. You can probably still find many of these in the Middle East, though the type mainly available in the West is a hazelnut-sized golden-beige colour.

The chick pea has now become the chief pulse crop of India, where it is called the Bengal gram. A pod will contain not more than 2 or 3 seeds. It is cultivated in all the Middle Eastern countries and still is a staple crop of Egypt and the eastern Mediterranean, where it is used in a number of dishes.

We buy chick peas dried in supermarkets, but it is well worth purchasing them in Italy, southern France, Spain and Mexico, where they are also cultivated as the new crop needs very little cooking and is a far finer flavour. The alternative name of garbanzo bean has Hispanic origins.

Perhaps the most imaginative and delicious recipes come from India. Julie Sahni tells us that the most famous dishes come from the Punjab, where the chick peas are cooked, 'either by braising them in tangy spicy cumin- and coriander-scented sauces or by boiling them and combining them with fresh herbs and seasonings, making delicious appetizers. In the south they are turned into a snack with an intriguing taste by adding jaggery (raw sugar) and coconut. Chick peas are also ground into a flour called *besan*, which forms the basis of batters for fritters, dumplings and sweetmeats and, of course, spicy breads.'

SPINACH AND CHICK PEA SOUP

Many Middle Eastern recipes using chick peas were brought to Spain by the Moors. I have enjoyed this Lenten dish many times in Spain; it is rather more than a soup and makes an excellent lunch.

75 ml / 5 tbsp olive oil
1 head garlic cloves, peeled
1 onion, sliced
225 g / 8 oz chick peas, soaked for an hour
450 g / 1 lb spinach leaves, chopped
2 large potatoes, sliced
sea salt and freshly ground black pepper

Heat the olive oil in a large saucepan, throw in the garlic, onion and drained chick peas and cook in the oil for a few minutes before adding 2.25 litres / 2 qt water. Bring to the boil and simmer for an hour.

Add the spinach and the potatoes and continue to cook for another half-hour. Season liberally and serve.

CHICK PEAS WITH YOGURT AND PITTA

Fattet Hummus

As chick peas are so redolent still of Egyptian cuisine, I have included a slightly simplified version of Claudia Roden's recipe for a famous chick pea dish, favouring the Damascus version, though not completely.

575 ml / 1 pt yogurt
170 g / 6 oz chick peas, soaked for an hour
150 ml / ¼ pt olive oil
3 garlic cloves, crushed
juice and zest of 1 lemon
large bunch of mint, finely chopped
2 pitta breads
3 tbsp pine nuts, toasted
sea salt and freshly ground black pepper

Pour the yogurt into a sieve so that it drains. This takes an hour, or you can do it overnight.

Boil the chick peas for an hour or until tender, pour half into a blender with a little of their liquid, the oil, garlic and lemon zest and juice. Liquidize to a smooth purée, season, stir in the mint and reserve.

Cut the pitta breads in two and bake in a hot oven until crisp. Leave to cool, then break up into bits and throw into a large bowl. Pour some of the chick pea water over the bread so that it softens, then cover with the drained chick peas. Pour over the purée of chick peas, then cover that with the strained yogurt. Sprinkle the top with the toasted pine nuts and serve.

HUMMUS

170 g / 6 oz chick peas, soaked
2 garlic cloves, crushed
juice and zest of 1 lemon
50–75 ml / 2–3 fl oz olive oil
handful of mint, finely chopped, or 3 tbsp tahini
sea salt and freshly ground black pepper

Boil the chick peas until they break easily on the point of a knife – about an hour, then drain reserving the cooking liquid. Place the cooked chick peas in a blender with the garlic, lemon zest and juice, seasoning and oil, liquidize, then add enough of the cooking water to make a thick, creamy purée.

If you want fresh-tasting hummus, add the mint; if you would like one with a darker, more aromatic flavour, add the tahini. Or you could add neither: chick peas alone have a wonderful earthy resonance that is very satisfying.

LENTIL

Lens culinaris

Among the first plants to be domesticated, lentils were daily fare in the Ancient World. The crops are still grown in the Near East and the Mediterranean regions, hence the misleading assumption that lentils can only grow in a hot climate. This is not true; lentils can be grown as far north as Britain and in my opinion should be – they need a light, warm soil. The plant resembles a vetch, is small (rarely taller than 50 cm / 20 in) and has short, broad pods. There are many varieties – brown, green, yellow, orange and red. They are used extensively in Indian cooking, *masoor* being popular with the Muslim community, but in all over 50 varieties are grown there. One Parsee dish called *dhansak* requires a mixture of from three to nine different kinds of lentils for its effect.

Like other common vegetables, lentils have at times been scorned by the upper classes. Aristophanes in his play *Plutus* speaks of a man who has just acquired riches as not liking lentils any more. Apicius includes several recipes for lentils, one being a barley soup enriched with lentils, peas and chick peas. But a sign of their popularity in ancient Rome is that the obelisk which stands today in front of St Peter's made the voyage from Egypt buried within 2,880,000 Roman pounds of small red Egyptian lentils.

Now, of course, lentils are very much back in fashion, but they have been spurned throughout history, beginning with the 'mess of pottage' with which Jacob purchased Esau's birthright. For this traditionally has always been considered to have been a lentil soup. Indeed, in Lebanon a dish of lentils, rice and onions called *migeddrab* is also known as 'Esau's dish of lentils'. I must admit that I love all types of lentils, for their earthy flavours. There is no doubt that a meal with lentils as a part of it is one of the most rewarding and satisfying.

NUTRITION

Of all the dried pulses, with the exception of the soy bean, lentils contain the most protein – 25 per cent. They are 54 per cent carbohydrate, and contain vitamins A and B, iron and calcium.

PREPARING AND COOKING

Lentils need no soaking; they can be cooked straight from the packet; but do watch them, for they absorb astonishing amounts of water, so make sure they are boiled in copious amounts. Various types need different cooking times. Choose the kind of lentil for the particular dish, and gauge cooking time accordingly.

All lentils make wonderful soups, especially good when the days start to get colder. The recipes are all very simple; little need be added, just garlic or onion, salt, pepper and a good quality oil, yet they are immensely satisfying. There is one, though, which has the addition of potatoes and greens (see below). All of these soups can be flavoured with spices or left plain so that the strong earthy taste of the lentil comes through.

VARIETIES

SMALL YELLOW OR ORANGE LENTILS: These give you the traditional Indian dhal or a purée for a lentil soup. They will cook down within 10–15 minutes into a mush. If you leave them to cool in what little water is left, they will absorb that too. If you place the purée in a blender, it will become very smooth and creamy. This purée tastes very fine without any other addition, except salt and perhaps a little butter or oil, but it can also be highly flavoured by all manner of Indian spices to make many types of dhal. One can also mix an egg into the dry purée, adding spices and herbs, chopped onion or garlic, and mould into cakes to fry as small patties or balls.

LARGE BROWN OR GREEN LENTILS: These are much used in Indian cooking. They need 10 minutes' cooking. At this stage they have a certain resistance and bite to them and they are best used cold as lentil salads with the addition of oil and chopped herbs. If you give them another 5 minutes' cooking, they will have softened a little more and be suitable for all sorts of purposes, as an accompanying vegetable or as part of a main casserole or main dish. The great thing about these lentils is that they can stay cooking for up to an hour and though, by that time, they will be very soft, they will not finally disintegrate and each lentil remains separate.

PUY LENTILS: The French *lentilles vertes du Puy* are considered the best and are at the moment in gourmet fashion. Small, green and delicate in flavour, they are, indeed, very good. They are also, because of their reputation, rather expensive. In all honesty, I cannot think they are so far superior to any other small green lentil that they earn their expense. Nevertheless, it is a matter of personal choice. They will be cooked within 15 minutes and are generally eaten alone, hot or cold, so as to enjoy their individuality. Garnish them with fried, caramelized onion rings and dress with walnut or hazelnut oil and lemon juice.

POTATO AND LENTIL SOUP

for 6 to 8

1 tbsp sesame oil
2 tbsp olive oil
1 onion
3 garlic cloves, chopped
115 g / 4 oz orange lentils
115 g / 4 oz brown lentils
450 g / 1 lb potatoes, peeled and chopped
115 g / 4 oz spinach, kale or spring cabbage, washed and sliced
sea salt and freshly ground black pepper

Heat the oil in a pan and throw in the orange lentils. Stir over a heat for a second, then add 850 ml / 1½ pt water, bring to the boil and simmer for 10–15 minutes, until they are soft. Leave to cool, then blend.

Meanwhile, in another pan, heat the oil and throw in the onions and garlic, fry for a moment before adding the brown lentils and potatoes. Fry for another second or two, then add 1.5 litres / 2½ pt of water. Bring to the boil and simmer for 20 minutes.

Halfway through the cooking time, add the chopped greens, stir them in and place the lid back on the pan. When these are cooked combine the two lentil mixtures, stir well and season. If the mixture is too thick, add more water or stock.

LENTIL BALLS WITH SPICY SAUCE

225 g / 8 oz brown lentils, cooked for 20 minutes, then drained and mashed coarsely
2 tbsp gram flour, sifted
2 garlic cloves, crushed
1 tbsp garam masala
½ tsp sea salt
little flour
1 beaten egg
3 tbsp toasted breadcrumbs
for the sauce:
250 g / 8 oz canned tomatoes
1 tbsp Dijon mustard
1 tbsp soy sauce
½ tsp Tabasco sauce

Add the gram flour, garlic, garam masala and salt to the mashed lentils, then fashion the mixture into balls. Refrigerate for an hour.

Roll in flour, egg and toasted breadcrumbs, then deep- or shallow-fry in vegetable or groundnut oil, so that the outside is crisp.

To make the sauce: blend the chopped tomatoes to a thin sauce. In a pan bring this to the boil and reduce it by a third. Add the rest of the ingredients, until you have a dipping sauce that pleases.

PUY LENTILS WITH CREAM

This is an excellent dish whether hot or cold.

2 tbsp olive oil
2 onions, chopped
225 g / 8 oz Puy lentils
300 ml / ½ pt double cream
sea salt and freshly ground black pepper
generous handful of mint, chopped

Heat the oil in a pan, throw in the chopped onions and cook for a moment, then add the lentils. Add 1.5 litres / 2½ pt water, bring to the boil and simmer for 20 minutes. Watch that the lentils do not soak up all the water (add more if you need to), then drain thoroughly. Pour in the cream, seasoning and chopped mint and serve.

SOYBEAN or SOYA BEAN

Glycine max

It is ironic that a bean which is so high in protein and nutritional richness, which could feed the hungry world ten times over, is such a dull and tasteless creature. Of all the fascinating beans, this is the only one that lacks real character and flavour. There are over a thousand varieties of soy – which is not a bean at all; it belongs to the pea family – and in such a range there ought to be one or two that are delicious. The problem is finding them.

The true worth of the soya bean, in my opinion, is in its by-products – flour, oil and milk. The last I find highly refreshing to drink in the summer, sometimes mixed half and half with yogurt, like a lassi. The oil is an excellent all-purpose polyunsaturated oil, while the flour can be added to breads, cakes and pastries for extra food value. The flour has become valuable commercially, appearing in ice-cream and many other food products.

The soybeans stocked in wholefood shops are usually the round beige-coloured type, or possibly the small black bean. There is little difference in taste between the two. Both varieties need soaking for around 10 hours in cold water, though if you pour boiling water over the beans, they need little more than an hour. A bean is soaked completely if it will split into two and the insides are flat rather than concave. If soaked for too long, goodness is lost in the water. Drain the soaking water from the beans and use new water for the cooking.

Another irony is that the soybean, unlike other pulses, has a substance called a 'trypsin inhibitor', which blocks a trypsin enzyme essential for the digestion of protein. The key is for them to be thoroughly cooked, then the trypsin inhibitor is destroyed. If simmered on top of the stove, soybeans will need around 4 to 5 hours. One of the best methods is to cook them overnight in an Aga or a Rayburn, or in an electric casserole. They can also be pressure-cooked for 25 minutes.

This is the science, but so far we are nowhere near the art. The main problem is that if the bean is left whole while cooking, it refuses with steadfast obstinacy to soak up the flavours around it. Hence, you have the insulated bland bean sitting amidst a sea of sauce. Breaking up the bean at the end of cooking helps. One whirl in the liquidizer will crush the bean into several bits, or a little hard work with the potato masher will do the trick too. Yet the bean bits still taste of nothing very much.

When the beans are pulped – blend the beans for a couple of moments until you have thick grainy purée – they can at last be flavoured. The beans can also be pulped after soaking, and the pulp cooked, when the cooking time ought to be half that for the whole beans.

ABOVE A selection of pulses

The cooked pulp can be mixed with chopped vegetables and spoonfuls of this mixture can be rolled in breadcrumbs, sesame, poppy or mustard seeds, and then fried. If the mixture is too liquid, a tablespoon of soya flour will thicken it; or, if it needs binding, add a beaten egg. I would tend to use strong flavourings - shallots, onions, garlic and herbs. Served with a miso sauce, these soybean croquettes can be far more delicious than they sound.

In America you can buy packets of dry roasted soybeans, salted like peanuts. While the nuts have 27 per cent protein, the beans have 20 per cent more. Packets come in plain, salted, garlic or barbecue flavour. You can roast your own at home. Using an oiled baking tray, spread the soaked, drained beans over the tray and place in an oven preheated to 180°C/350°F/gas4 for about 2½ hours, or until they are dark golden and crisp. Shake them twice, at the end of each hour. Toss the roasted beans in a flavoured salt – celery, garlic or sesame – or in your own flavouring made out of herbs and spices.

If the roasted soybeans are left unsalted, you can make 'kinako'. Grind the roasted beans, or blend them in a liquidizer, until they are a powder. This flour has no gluten so it will not thicken, but it can be used to coat food before frying. It can also be mixed with butter or margarine, miso or ground peanut or sesame, to make a paste for spreading on toast and biscuits. In Japan they sweeten kinako and use it to coat confectionery. It must be about the only example of chewy sweets which could be good for you.

The recipes which follow on the opposite page are for the whole bean, an attempt at solving the challenge of their insipid flavour. The first recipe is an alternative to roasting the beans.

SESAME SOYBEANS

100 g / 7 oz soybeans
sunflower oil for frying
1 tbsp sesame salt (that is, 1 part of sea salt to 5 parts roasted ground sesame seeds)

Soak the beans, drain them, then lay them out to dry on paper towels.

Heat 1 cm / ½ in sunflower oil in a deep pan, throw in the beans, fry them for about 10 to 12 minutes. They are done when medium brown and crisp.

Let them drain on paper towels and, when cool, mix in sesame salt. They will keep for months, if they are in an airtight container.

SOYBEAN FRITTERS

200 g / 7 oz soybeans, cooked
200 g / 7 oz brown rice, cooked
4 small onions, chopped
2 garlic cloves, crushed
1 tsp each ground rosemary, dill seed, marjoram, celery salt
pinch of cayenne pepper or chilli powder
1 egg
1½ tbsp soya flour
sunflower oil for frying
few drops of sesame oil

Mix the ingredients well together. Shallow-fry spoonfuls of this mixture in sunflower oil which has been flavoured with a few drops of sesame oil. Cook until crisp on both sides.

ADJUKI BEAN

Phaseolus angularis

This is one of the beans used for sprouting, but it can also be soaked and cooked. It is small and red, highly popular and cultivated widely in China and Japan. In the latter country it is boiled, mashed and sweetened, then used extensively in cakes, sweets and red rice. The beans are even sold powdered.

Nutritionally they are rich, with 25 per cent protein, and a good source of vitamin C and many of the vitamin B complex.

MUNG BEAN

Phaseolus aureus

This is the tiny green bean most commonly used for sprouting. When we buy or eat bean sprouts in the West, we are usually eating the mung bean. The sprout is particularly rich in protein – 35 per cent – and also has B and C vitamins. B1 (thiamine) goes on doubling each day until day four. In China and India the bean is soaked as well as sprouted. They do not need soaking and will cook quickly within 20 minutes. It is eaten in India in many of the *dhal* dishes. I include one here.

MUNG DHAL WITH YOGURT

3 tbsp sunflower oil
1 tsp turmeric
½ tsp fenugreek seeds
½ tsp cumin seeds
½ tsp mustard seeds
½ tsp asafoetida
2 green chillies, chopped
170 g / 6 oz mung beans
150 ml / 5 fl oz plain yogurt
1 onion, finely chopped
sea salt and freshly ground black pepper

Heat the sunflower oil in a large pan and throw in all the spices. Cook them until they begin to roast and expel their oils.

Add the asafoetida and the green chillies, cook for another second or so, then add the mung beans. Stir thoroughly, then add 1.5 litres / 2½ pt of water. Bring to the boil and simmer for about 20 minutes, leave to cool.

Drain any excess liquid, though the beans should have soaked up all the liquid. Add the yogurt and onion and season. Serve warm.

MUNG BEAN *Phaseolus aureus*

ALFALFA

Medicago sativa

We are familiar with this member of the pea family only as a seed for sprouting, but I include it as it is a kind of miracle food. Also known as lucerne or buffalo herb, the plant has extraordinarily long roots, which can easily grow to some 13 m/40 ft beneath the ground. Sprouted alfalfa seeds have a high protein content (40 per cent) – higher than beef or milk. They contain vitamins A, B and C, with good amounts of D, E, G, K and U. They also contain B12, calcium, iron, sodium, potassium, sulphur, phosphorus, silicone, aluminium and magnesium. Deep-rooted plants like this that can reach into the subsoil gain minerals not available to shallower-growing plants. Alfalfa could solve the Third World's malnutrition problems, but although alfalfa powder is sent to developing countries, it has not gained wide acceptance.

SPROUTING SEEDS

All manner of vegetable and cereal seeds can be sprouted, but members of the bean family are among the stalwarts. Mung bean sprouts, sold as Chinese bean sprouts, contain 37 per cent protein, plenty of vitamin C and many of the B complex vitamins. Adjuki beans are equally vitamin-rich, with 25 per cent protein. Growing sprouts contain vitamin C (ascorbic acid is always concentrated in rapidly growing tissue). Other vitamins and minerals soon appear. After five or six days' germination, sprouts also contain vitamins from the B complex group, including B12 and vitamins E, G, K and U.

Sprouts taste excellent and are undeserving of their minor place on our tables, relegated to the fringes of vegan- or vegetarianism, or macrobiotic diets. Try experimenting using sprouts as a major ingredient in sandwiches, where their crunchy texture can be appreciated. Mix with a sharp spread like anchovy, a yeast extract mixed in with slivers of onion, or avocado and rocket. They are delicious eaten just as they are, with a little sea salt, lemon juice – and maybe walnut oil. But they are so refreshing to the palate they really need no dressing at all.

LEFT top: Mung dhal with yogurt (previous page); bottom: Sesame soybeans (previous page)

Salads made from sprouts are a gift to slimmers. Tasty and with plenty of crunch, they are filling, need minimal dressing, and provide many of the vitamins and trace elements necessary to the metabolism. Or stir-fry, which does not destroy sprouts' nutritional value. Throw them in at the last minute, with perhaps a few courgette slices, in a little sesame oil, garlic and ginger. They will only need half a minute's stir-frying. They should still have plenty of crunch.

Bean, alfalfa and onion sprouts can be bought ready-packaged at health food shops and supermarkets. Making your own is, of course, a lot cheaper. Warmth and water are needed to activate the seed's dormancy into growth. It is best to splash out initially for a bean-sprouter, available at health food stores. This consists of three perforated trays which drain into a bowl beneath. You sprinkle seeds in the trays and pour cold water through them each day. The idea is to keep the seeds moist. You can improvise a sprouter from a jar with a net covering, but do make sure you drain all the water, so the seeds never lie in a puddle. Seeds will sprout in darkness or light, but dislike direct sunlight and cold: the optimum temperature is 13–21°C/55–70°F. Most take between three and six days. Once the seeds have sprouted, keep them in a covered container in the cool or in the refrigerator for up to a week.

A word of warning. Always buy seeds intended for sprouting: health food shops generally have a range. Seeds from other sources intended for planting may have been treated with fungicides or insecticides.

- Alfalfa: High protein and vitamin content. Rinse once daily. Sprouts in about five days.
- Adjuki bean: High protein and vitamin content. Rinse once daily. Sprouts within four days.
- Lentil: Sprouts in about five days.
- Mung bean: Soak in water for 12 hours until the husk splits, then transfer to the sprouter. Sprouts in four days. Before using, rinse off husks in cold water.
- Soybean: Soak before sprouting, as for mung bean. Rinse four or five times daily. Sprouts in three to four days. Needs to be cooked before eating.

Solanaceae

THE POTATO FAMILY

POTATO *Solanum tuberosum*
TOMATO *Lycopersicon esculentum*
AUBERGINE *Solanum melongena*
SWEET PEPPER *Capsicum annuum* Grossum Group
CHILLI PEPPER *Capsicum annum* Longum Group
and *C. frutescens*

What an astonishing change occurred in the Old World's diet after key members of this family were introduced from South America. Imagine the food of the twentieth century without the conspicuous bulk of the potato, the ubiquitous flavour of the tomato or the fiery power of the chilli pepper. Yet the change in most cases was far from abrupt.

The group's alternative botanical name is the nightshade family, and unfamiliar members were often treated with justifiable suspicion. Some powerful poisons run in the family's veins. The ancients were familiar with belladonna, henbane and mandrake; a newcomer arrived on the scene with the discovery of nicotine, from tobacco. Even the everyday food plants such as tomato and potato contain toxic alkaloids in their leaves and stems. (No wonder some of our ancestors chose to grow them for decoration only.)

Today we still grow solanums and nicotianas as garden ornaments along with our petunias and daturas, as well as getting visual delight from the bright fruits of tomatoes and peppers.

POTATO

Solanum tuberosum

The potato has become one of our major staple foods. It is the world's fourth largest crop and 90 per cent of this is grown in Europe. Its rise to this eminent status has, however, been a slow and chequered one, constantly dogged by a poor press and confused by misunderstandings. The potato is a native of southern Chile, growing high in the Andes around Lake Titicaca. The earliest sign of its consumption can be dated back to 3000 BC, and it became the Incas' staple food, but it remained a secret from the rest of the world until 1553, when Pedro de Leon wrote about it in the *Chronica del Peru*. The *conquistadores* brought the potato back to Spain, but no one took much interest in it. However, it was found to grow well in the northern province of Galicia, where the climate suited it, and began to be used by the Spanish as a food for the poor and the sick, and as army rations.

The introduction of the potato to the English-speaking world has been inextricably linked with Sir Walter Raleigh. It was he, we were always taught, who brought the potato from Virginia to England in 1586. But the potato was unknown in the whole of North America and Mexico until the eighteenth century. What the English adventurers brought home was not from Virginia, and nor was Raleigh responsible for its presence on the ship. A large supply of tubers formed part of the provisions taken on board at Cartagena, in Colombia, by Sir Francis Drake as he sailed homeward after harassing the Spaniards in the Caribbean. On his way he called in at Virginia to collect the disgruntled settlers – among them Thomas Hariot, one of Raleigh's men, who is thought to have taken an interest in the unusual foodstuff. Back in England, the new potato is supposed to have been distributed to interested parties, including Raleigh, the herbalist John Gerard and even the Queen. There are accounts of people mistakenly eating the leaves or fruits and throwing away the edible tubers, thus contributing to the low culinary esteem in which the potato was held.

Some people assumed that what Drake brought back must have been the sweet potato, since this is the kind that could be grown in the Caribbean where he provisioned his ship and the one that seems to have been more widely known. Whatever the precise nature of the tubers Drake imported, the spud we know must have arrived in England around that time. Both types of potato were evidently current in London by the 1590s. Within a decade of Drake's return both were being discussed by Gerard in his *Herball*. Unfortunately Gerard fell into – and perpetuated – the error that since Drake's last landfall had been Virginia, that region was the source of the newer kind of potato. Gerard and his contemporaries were obviously already familiar with the sweet potato, of which he had 'planted divers roots (which I bought at the Exchange in London)' but which 'perished and rotted' in the winter cold. Indeed, the sweet potato was 'the common Potato' to which Gerard compared the newer 'Virginian Potatoes'. Gerard's opinion was a lot kinder than those of many later writers. He described the new arrival as being similar in many ways to the sweet kind, '...equal in goodnesse and wholesomenesse to the same, being either rosted in the embers, or boiled and eaten with oile, vinegar and pepper, or dressed some other way by the hand of a skilfull Cooke.'

The potato colonized the world by devious routes. As it originated in South America, you might expect it simply to have travelled northwards – and perhaps to have arrived in Virginia for the first settlers to encounter. But, as we have seen, the potato touched Virginia only briefly, in the ship by which the settlers were leaving. As a serious crop it reached North America only in 1720, when Ulstermen settled in New Hampshire, bringing their Irish potatoes with them. There it was quickly renamed the Murphy, and by 1760 was being grown commercially. So how had it got to Ireland?

One story goes that potatoes were introduced when they were planted at Raleigh's property at Youghal, near Cork, in 1586. (If the provisions brought back by Drake had indeed consisted only of sweet potatoes, they would have proved short-lived, even in the mild winters of southern Ireland, but we know that by the early seventeenth century the white potato had become a staple food.) A more fanciful story is that potatoes formed part of the stores of the Spanish Armada and turned up in wrecks on the Irish coasts. Being miraculously hardy, the seed potato could easily have survived shipwreck and drowning; as the Spanish fighting forces (with the poor and the sick) were fed on potatoes, there would certainly have been plenty on board. Fanciful conjectures have a habit of turning out to be true, and I would put my money on that explanation. A more prosaic account is that it was introduced through trade with Flanders.

For by the last quarter of the sixteenth century, the potato had arrived in northern Europe. It took root in many places where cereals were difficult to grow, and was cultivated by poorer people to supplement their own diet or to feed to animals. Potatoes grown in Galicia were shipped to Genoa and were eaten by the poor in northern Italy. From there they moved to Switzerland and on to Antwerp, where a specimen was painted for the first time. (Gerard used this illustration in his *Herball*, although the frontispiece portrait depicts him holding the sweet potato.) It was in Switzerland in 1598 that a woman – bless her – actually gave a recipe for roast potatoes. Who, then, would be the first to suggest mashing and puréeing the divine spud with cream and butter?

No one yet: they were all too busy criticizing it. The potato has never been an overnight success. Experts were only too ready to decry it as tasteless while commending it as food for the poor. As early as 1663 the Royal Society recommended growing potatoes as a safeguard against possible famine, but few heeded. John Forster, apparently an ardent fan, published *England's Happiness Increased, or a rare and easie Remedy against all succeeding Dear Years: by a Plantation of the Roots called Potatoes*. But still the potato did not catch on, even among the people worst affected by the 'dear years'. Disturbed by a succession of cereal crop failures, in 1720 Frederick William I commanded the Prussian peasants to plant potatoes. There was widespread resistance to the new crop; the peasants pulled the plants up again, fearing that eating potatoes gave people leprosy. They were only reassured by seeing Frederick William himself eating potatoes on his palace balcony. (How were they served, one wonders – roasted, sautéed or mashed? Alas, the observer does not tell us.)

Waverley Root in his food dictionary points out that people's reluctance to eat potatoes in the belief that they were harmful or poisonous had some justification. Solanine, which shows up as green patches on the potato (and which we cut out when we find it) is a poison capable of causing an unpleasant stomach upset. Root claims that the solanine in early potatoes – before they had been selectively bred – was of a higher strength and caused skin rashes which frightened people into believing they had leprosy.

Such a bad press hampered the potato from becoming a large commercial crop. A botanist in 1719 writes grudgingly, 'they are not without their admirers, so I will not pass them by in silence.' Another botanist, Philip Miller of the Chelsea Physic Garden, writes in 1754: 'they are despised by the rich and deemed only proper food for the meaner sort of persons.' The French naturalist Raoul Combes writes in 1749, 'Here is the worst of all vegetables...', adding with a sneer, '...nevertheless the people, which is the most numerous part of humanity, feed themselves with it.' Brillat-Savarin appreciated the potato only as a protection against famine, dismissing it as 'eminently tasteless'. Diderot shared his opinion: 'This root, no matter how you prepare it, is tasteless and floury.' Nietzsche concluded that just as a diet composed mainly of rice led to opium, so one of potato must lead to alcohol.

In Scotland the eating of potatoes was opposed by Presbyterian ministers, who considered them 'an ungodly food because they are not mentioned in the Bible'. In England potatoes had become a field crop in the 1780s, but were mostly grown to feed cattle. But by 1832 the English authorities had recognized their nutritional worth and amended the Bread Act to allow potato flour to be used in the dough. In *The London Poor* Henry Mayhew writes of the street trade in baked potatoes in the 1860s: 'many gentlefolks buy them in the street and take them home for supper in their pockets; but the working class are the greatest purchasers. Many boys and girls lay out a half-penny on a baked potato.' Mayhew's distinction between different types of potato sounds familiar to modern ears: 'potatoes with a rough skin are selected from the others because they are the mealiest. A waxy potato shrivels in the baking.'

Perhaps the change of heart in England had been helped by the official popularization of the potato in France during the late eighteenth century. A sponsored contest for 'the best study of food substances capable of reducing the calamities of famine' had been won by Antoine Parmentier, a military pharmacist. Though we know Parmentier's name because it has been given to a soup, a hash and an omelette, all containing potatoes, in his lifetime he was best known as a writer and for introducing a vaccine against smallpox in the army. Parmentier is said to have served a dinner at which all the courses – soup, entrée, entremets, salad, cake, biscuits and bread – were made of potatoes.

The potato was well on the road to its current dietary importance. It was planted wherever the climate suited it, and proved a good crop for on land where it was difficult to cultivate cereals or leafy vegetables. It is a paradox that the crop intended to eliminate hunger was the cause of some of the nineteenth century's worst famines. Upon the poor soil of Ireland, the potato had flourished when cereals had failed, and provided six times the nourishment. Every small cottage garden contained its plot of potatoes. William Cobbett referred to the potato as 'Ireland's lazy root'. One of the results, according to Charles Edward Trevelyan, permanent head of the British Treasury at the time of the famine, was that Irish women forgot how to cook. 'There is scarcely a woman of the peasant class in the West of Ireland whose culinary art exceeds the boiling of a potato. Bread is scarcely ever seen and the oven is unknown.' When blight destroyed the harvest in the 1840s, even the potato disappeared; up to a million Irish died of starvation or the diseases that followed.

The taste for the potato survived the lean years. Advances in understanding the nature of the diseases and improved cultivation techniques gradually assured more reliable potato harvests. Whether plainly boiled to supply bulk, transformed into a culinary masterpiece or processed into chips and crisps, the potato's place in the twentieth century is undisputed. In Europe and North America most people probably eat potatoes on average once a day.

NUTRITION

Potatoes are 80 per cent water and 2 per cent protein, but are high in vitamin C and potassium. The nutrients tend to be in the skin and just beneath it. Potatoes also contain iron, phosphorus, calcium, sodium, sulphur and vitamin B. No wonder they have become a staple food for the poor. The more a potato is stored throughout the winter and the more a potato is processed, the less nutrition it will have. So eat your potatoes as soon as you can after they are dug, and eat them without peeling if possible.

VARIETIES

There are over 400 varieties of potato. If you are a gardener you will take pleasure in growing the potatoes you enjoy but cannot purchase commercially. I love growing the waxy salad varieties: La Ratte, Pink Fir Apple, Charlotte and Belle de Fontenay. Of these four, the first two have more flavour (but do remember that your soil and its particular combination of minerals is what gives a potato its own idiosyncratic taste). All of these potatoes are small and yellow-fleshed, perfect for summer salads or for boiling. One of the greatest potatoes for baking I have ever discovered is Golden Wonder; simply baked, it has a heavenly flavour. The crop is mainly grown in Scotland and Northern Ireland and goes straight into processing for crisps. When I tried to get seed potatoes to grow them in my own garden I was told firmly that the flavour I was so enthusiastic about came from northern Celtic soil and could not be reproduced in East Anglia. Perhaps. So far I have not had the opportunity to find seed potatoes to test the hypothesis.

Potatoes are divided up commercially into Earlies and Main-crop. One of the most famous early potatoes is Jersey Royal, a small, kidney-shaped tuber which should be washed but not peeled, for the potato is always harvested before the skin has set. In Jersey they proudly say that the potato cannot grow anywhere else, citing examples of attempts which always failed. The Jersey Royal appeared only a hundred years ago, when a Jersey farmer, Hugh de la

Haye, cut a large potato with fifteen eyes and planted each eye. The result was a crop of these tiny potatoes. Their flavour was unique and the crop was nurtured and valued for its early growth. I was in Jersey in January as the seed potatoes were being planted. Every scrap of land is carefully utilized, even the smallest strip and often on the steepest slopes. The end product can hardly be bettered for flavour.

At a potato testing at Le Meridien Hotel in London's Piccadilly, attended by many food writers, there was general agreement on the superiority of those potatoes which were grown organically. I have long thought that it is in the root vegetables that organic agriculture appears to show a marked improvement in flavour. Those standbys of the British housewife, the King Edward and Desirée, tasted dull and insipid even at Le Meridien; the organically grown potatoes were by far superior, having a full earthiness which was immensely satisfying. At this event we tasted the rare but beautiful Purple Congo. These blue potatoes stay blue after cooking; the colour does not leach out into the water. I admit they look a little odd on the plate. The Purple Congo spud is actually dark violet and distinctly reminiscent of Elizabeth Taylor's eyes. The other shock about the blue potato is that the flavour is not markedly different from the King Edward.

Other early potatoes include Arran Comet, Maris Peer, Pentland Javelin and Wilja. Maincrop potatoes include Desirée, Maris Piper, Pentland Crown and King Edward.

CHOOSING

It is perhaps most helpful for the cook to know which varieties are good for which purposes – for salads, for mashing, and for deep-frying or baking.

SALADS: Waxy varieties such as La Ratte, Pink Fir Apple, Charlotte, Belle de Fontenay, Jersey Royal.

BOILING OR STEAMING: All of the above, plus any Cypriot, Egyptian, Dutch or Belgian variety.

SAUTÉING: All of the above two categories, as the potato has to be firm to keep the shape.

MASHING INTO A PURÉE: King Edward, Golden Wonder, Wilja, Maris Piper, Ulster Sceptre and similarly floury types. Lindsey Bareham in her excellent book *In Praise of the Potato* comments that the *pommes purées* of Jöel Rebuchon or Frédy Giradet is *not* the mash we know and love. They use La Ratte with cream and olive oil and although this wonderful golden waxy spud gives a flavour resonant with 'new potato' taste the texture can never be as fluffy as a mash made with floury potatoes.

ROASTING: King Edward, Maris Piper, Pentland Dell, Desirée and Cara. Almost any potato will roast satisfactorily, but the floury ones, above, give slightly better results. Initial parboiling makes the exterior ragged and this, when roasted, soaks up the fat unevenly and crisps with the sea salt beautifully. Lindsey Bareham selects Pentland Dell as being by far the best at this.

DEEP-FRYING: Maris Peer, Ulster Sceptre, Desirée and King Edward.

BAKING: The floury varieties are best, so use all those potatoes which are good for mashing, especially Golden Wonder, if you can ever find them.

STORING

New potatoes are best dug or bought little and often, as you need them. Avoid buying bags of prewashed potatoes – they never keep so well. Main crop potatoes will store happily throughout the winter. If you have a garden, they can often be left in the earth and lifted late – before the first frosts. Store in the dark

and the cool. They will slowly lose their nutritional richness and to some extent their flavour, too. Potatoes which begin to sprout in the early spring can still be eaten if they are firm when gently squeezed. Simply take care to cut all the shoots out before cooking in whichever fashion you decide. The shoots are high in solanine, a poisonous alkaloid; this also makes the potato green, so cut out all green bits before using.

PREPARING

All new potatoes can be washed under running water and left unpeeled. Remember, the peel is an essential part of the nutriment. (One of the tragedies of the Irish is shown in a film of the 1920s where a family of farm workers is seen eating potatoes from a pile in the centre of the table. They are hot and steaming but each potato is peeled before being salted and eaten so most of the food value is lost.) Older potatoes will probably need scrubbing to get the earth and dirt removed. When you come to peel the potatoes, make sure you buy a peeler that is efficient enough to remove only a very thin layer of the outer skin. I find the swivel peeler is by far the best for this purpose.

For gratin dishes potatoes have to be cut in thin discs. Use a mandoline or a food processor, wash the potatoes in cold water to rinse some of the starch away, drain well and pat dry.

COOKING

BOILING: Put new potatoes into enough salted boiling water to cover and cook for 15 – 20 minutes. The timing depends upon the size, variety and how recently they have been dug up. I have known very small potatoes, no bigger than quails' eggs, just dug from my garden, being cooked within 10 minutes: how good they were, too. If you are using the potatoes for a salad, have the dressing ready in a bowl and add the drained hot potatoes to the dressing immediately. Hot into cold helps to fuse dressing and potato, and the potato absorbs more of the dressing.

BAKING: Pick out fairly large potatoes; the size of a mango is about right. Wash them by scrubbing under cold running water, then prick the outsides with a fork to stop them from bursting inside your oven. Cook in the oven at 160°C/325°F/gas3 for 2 – 3 hours, or 190°C/375°F/gas5 for 1 hour. Squeeze the crisp outside very gently: if it gives inside, they are done. Cut them in half and serve with butter, sour cream, grated cheese or any favourite sauce. At this stage the insides can be scooped out and mixed with additions, then restuffed into the potato skin. The emptied skins can also be sliced, fried in olive oil and eaten as an appetizer. Half a baked potato can be used as a container for other additions too, if some of the potato is removed: two poached or fried eggs with grated Parmesan cheese, anchovies and sour cream, or chopped onion with grated Cheddar cheese. They offer a host of possibilities for the inventive cook and make simple supper dishes as well as good party fare.

ROASTING: Peel the potatoes and cut them into pieces about the size of half an apricot or peach. Drop these into boiling water and let them cook for 3 minutes. Drain them thoroughly. Drop them into the fat or oil surrounding roast meat, fowl or game, sprinkle them with a little salt and roast for about an hour, turning them once so that they are golden brown all over.

STEAMING: An excellent way with new potatoes is to lay a bed of mint leaves at the bottom of a steamer and place the washed new potatoes on top. Steam for 15 to 20 minutes. If making a potato salad, this is a perfect way to cook the potatoes beforehand.

MASHING: Boil the potatoes until tender, then drain. Place the cooked potatoes back in the saucepan over a very low heat to drive off any excess moisture. Turn off the heat. Pour in a little milk and a few knobs of butter. Then use a potato masher to combine everything, and season with salt and pepper (the last is important). How much milk and butter you use is up to personal taste: potatoes will soak up great quantities of fat, which is why in the past they were always considered fattening. But potatoes also need very little to turn them into a smooth purée. Instead of milk, you can use cream for a richer effect, fromage frais for a very slightly cheesy tang and a good texture (perfect foil for a poached fish dish), sour cream,

RIGHT *Boiled pink fir apple and purple congo potatoes*

which goes well with a thick gamy stew, or add masses of chopped parsley, dill or coriander, depending on what dish you wish your mash to accompany.

Never, but never, use a food mixer to turn your potatoes into a purée. What happens is that you turn it into wallpaper glue. It goes grey, gooey and sticky.

However, the best-quality mashed potatoes are also whisked after being mashed. This merely allows the air in and gives you a smoother, fluffier, lighter mash altogether. Use a fork or a wire whisk, but remember that whisking only works if your additions are fairly light in themselves, hardly more complicated than milk and butter.

SAUTÉING: Peel and boil the potatoes for 15 minutes, then drain them thoroughly. Slice into thickish discs of 5 mm / ¼ in or a little less. Heat some olive oil or sunflower oil (the first gives you flavour, the second a lighter, crispier potato), then throw in the pieces of potato. Fry until golden brown and crisp, turning them to ensure that they are done on both sides.

DEEP-FRYING: Responsible cookery books warn you of its dangers. My mother once set the kitchen on fire and I have to confess that I have only ever attempted deep-frying about twice in my life – and using a wok. In my view it is not only dangerous, but wasteful; a normal domestic kitchen cannot re-use all that amount of oil. It is also unhealthy, for the chip absorbs vast amounts of fat, and frankly I believe the end result is hardly worth the bother. The reader will conclude that the chip is not a favourite food for me, any more than the commercial potato crisp.

My advice is – if you must have chips – shallow-fry. Make sure the chips are cut not too thickly. Rinse and dry thoroughly before cooking in hot olive oil or one of the polyunsaturated vegetable oils. Move the chips around the pan to make sure they are brown. They will cook within 10 to 12 minutes.

BAKING IN A BAG: A good method for smallish potatoes is to use a roasting bag, placing the potatoes in it with a little olive oil and some chopped mint and sea salt. Close the bag and give it a shake. Place in an oven preheated to 190°C/375°F/gas5 for 30 minutes.

FRYING IN A CLOSED SAUCEPAN: This method is excellent with new potatoes, and stems from a Richard Olney recipe, where the washed potatoes are placed in a pan that has a layer of olive oil in it. Then about 2 heads of garlic are added (the cloves are *not* peeled). Add a little sea salt and then put the pan over a low heat, place the lid on and leave. Just shake the pan occasionally. The potatoes are cooked within 40 minutes; the garlic is steamed.

I must point out that this recipe is only for garlic lovers. We once ate this for lunch, then that night crossed to France sharing our berth with a friend. In the morning, she commented that the cabin was so thick with garlic fumes she could hardly breathe.

However, with or without garlic, the method is a perfect one, for the potatoes fry and steam at the same time. I have made it using onions, shallots and leeks, all delicious.

GRATINS: In this range of recipes, the potato becomes ennobled, raised to aristocratic eminence by being baked in the oven in shallow earthenware gratin dishes. These classic dishes are by far my own favourites. The potatoes need to be peeled, then sliced very thinly either on a mandoline or in a food processor. Wash some of the starch away from the slices by soaking in cold water, then pat dry. This is important, otherwise the slices will adhere together into a gluey stodge; what you want are layers of well flavoured individual slices encased in a creamy sauce. There are many versions of each of the classic recipes. I give my preferences here, but every good French cookery book will have its own. It is worth trying them all to find the recipes that suit you.

• For Gratin Dauphinois: Sprinkle a little salt, black pepper and nutmeg between the layers, then pour 600 ml / 1 pt of cream over the potatoes. Some cooks suggest par-boiling the pieces in milk for 2 minutes to reduce the acidity of the potato slices, and prevent the cream from curdling. I have not found this necessary. Merely cook the gratin slowly, say for 2½ hours at 160°C/325°F/gas3.

• For Gratin Lyonnais: Interleave potato slices with thickly sliced onion, pour over enough meat or vegetable stock (or use cream) just to cover. Finish with Gruyère or Cantal cheese grated over the top.

• For Gratin Crécy: Interleave potato slices with thickly sliced carrot and use either stock or cream as above.

POTATOES SAUTÉED WITH ONION

I first learnt to cook this robust dish in the 1950s when I was living in Vienna.

675 g / 1½ lb potatoes
3–4 tbsp olive oil
2 large onions, sliced
sea salt
good handful of chives, chopped

Peel the potatoes, boil them for 15 minutes, then slice them into discs.

Heat the oil in a large frying pan, throw in the onions followed by the potatoes. Fry until both are golden and crisp, about 10–15 minutes

Serve sprinkled with sea salt and chives.

POTATO SALAD

900 g / 2 lb waxy new potatoes, unpeeled
2 red onions, thinly sliced
1 bunch of spring onions, chopped
good handful of parsley, chopped
2 tbsp capers
300 ml / ½ pt aïoli mayonnaise
for the dressing:
75 ml / 3 fl oz olive oil
zest and 1 tbsp juice from 1 lemon
1 garlic clove, crushed
sea salt and freshly ground black pepper

Boil the potatoes in lightly salted water. While the potatoes are boiling, make the dressing in a large bowl with the olive oil, lemon zest and juice, garlic and seasoning. Drop in the sliced red onions and let them marinate for a while.

When the potatoes are cooked, drain them, then slice into rounds and add to the dressing at once. (I never peel potatoes for a salad, because the peel adds so much flavour, but purists think the peel unsightly.)

When cool, add the chopped spring onions, parsley and capers, then fold in the aïoli mayonnaise. Refrigerate until eaten, but not for too long; it should be cool, not iced.

POTATOES WITH CELERY

900 g / 2 lb waxy potatoes
2 heads of celery
3–4 tbsp olive oil
30 g / 1 oz butter
1 tbsp celery salt
1 tbsp soy sauce
handful of parsley, finely chopped

Scrub the potatoes but do not peel them. Boil them for 12 minutes, then drain and dice into 1 cm / ½ in chunks. Clean the celery and chop the central heart into 1 cm / ½ in slices. Heat the oil and butter, throw in the vegetables, add the celery salt and fry until the potato is crisp and brown. Finally add the soy sauce and stir-fry for 3 more second, then sprinkle the parsley over and serve.

GRATED POTATO CAKES

900 g / 2 lb large potatoes
115 g / 4 oz cream cheese
55 g / 2 oz plain flour
55 g / 2 oz Parmesan cheese
55 g / 2 oz Gruyère cheese
1 onion, finely chopped
2 eggs
handful of parsley, finely chopped
150 ml / ¼ pt single cream
sea salt and freshly ground black pepper
vegetable oil for frying

Peel the potatoes and grate them. Place in a colander and put them under cold running water to get rid of the starch, turning them with your hand. Pat and squeeze them dry in a clean cloth.

Mash the cream cheese with the flour, cheeses, onion, seasoning, eggs and parsley. Stir in the cream and the potato and mix thoroughly.

Heat some oil in a pan and drop spoonfuls of the mixture into it; the cakes should be about 5 cm / 2 in wide (if too big, they will break). Cook 4 at a time, allowing them to get golden brown before turning them over and cooking the other side. Serve at once.

TOMATO

Lycopersicon esculentum

It is well-nigh impossible today to think of food without the tomato and its almost worldwide flavouring role. Who would have thought that its popularity could have grown so great, after the indifference and suspicion following its discovery? It took almost until the dawn of this century for the tomato to find acceptance. Its family, after all, contained the deadly nightshade, and at first people considered that this golden fruit must be either poisonous or an aphrodisiac. This explains its first name, *pomo d'oro* – 'golden apple' or 'love apple', for the earliest tomatoes to arrive here were the yellow variety, only now creeping back into our seed catalogues.

Gerard's *Herball* is most unflattering: 'the whole plant is of a ranke and stinking savour'. Gerard links it with a derogatory remark about the Spanish, England's enemy at the end of the sixteenth century, who we are told eat these 'apples' boiled with pepper, salt and oil but 'they yield very little nourishment to the body, and the same naught and corrupt'. Rather than being eaten, tomatoes were used instead as decorative climbing plants. Perhaps some unfortunate person had eaten the leaves or stalk instead of the fruit and suffered abdominal pains, which is one of the mistakes that gave the potato its unhappy reputation.

The tomato originally grew in the Lower Andes, an area that now covers Ecuador, Peru and Bolivia. The ancient Peruvians, it is thought, picked the small cherry-sized tomatoes in the wild and did not bother to cultivate them. But in its short season it was obviously enjoyed, because the tomato travelled northwards as a weed long before the coming of Columbus (see page 124), and was very likely cultivated in Mexico. Its name is derived through the Spanish *tomate* which, in turn, comes from the Aztec *tomatl*.

We do not know how the tomato reached Italy so soon – in 1522; since the Kingdom of Naples was under Spanish rule, it seems likely that all the foods from the New World were soon introduced to southern Italy. They crept into southern France, but most amazingly they were also soon known in Poland – their Polish name was *pomodory*, almost the same word as in Italian. In England a few herbalists began to grow tomatoes in their private gardens as an ornamental curiosity. Olivier de Serres, agronomist under the French Henri IV, wrote 'they serve commonly to cover outhouses and arbours.' It was not until the eighteenth century that two Italian Jesuit priests returned with the red tomato variety, which quickly ousted the yellow in popularity. The Italians and the Spanish were the first to enjoy tomatoes as a salad, a fact which caused some astonishment elsewhere. Philip Miller, in charge of the Chelsea Physic Garden, tells us in 1752 that Italians and Spaniards ate them like cucumbers with pepper, oil and salt and some even ate tomatoes in sauces and as soups. Miller adds that the tomato gives 'an agreeable acid to the soup' but he thinks that the nourishment they afford 'must be bad'. Poor tomato.

Reactions in America were, if anything, worse. Long cooking before the tomatoes could possibly be palatable was earnestly advised. 'Tomatoes will not lose their raw taste in less than three hours' cooking', Eliza Leslie wrote in 1848. The vinegar and spices added to tomato sauce were considered to be some antidote to any lingering poison. The sauce crept into English cooking at the beginning of the nineteenth century, for Pickwick is demanding chops with tomato sauce in 1836. The fact that the tomato survived at all must be due to the enthusiasm that amateur gardeners had for the vegetable (and still have, for that matter), so that rural cookery of the nineteenth century was full of tomato chutneys, pickles and relishes.

Now tomatoes are a huge commercial crop. Canned, whole or chopped, made into juice, condensed into soup, used as flavouring for baked beans and numerous other convenience foods, they are one of the world's best-selling foods. Yet served as a raw salad vegetable, their quality has declined into a woolly, hard-skinned, tasteless ball of innocuous pap. Raw tomatoes are bitterly criticized, and have been for decades. Food producers and commercial growers are aware of public reaction, and continuous research goes on to find a tasty tomato with shelf life. There are some rays of hope. The little cherry tomato can be so acid that it sometimes seems like chemical war-

fare upon the palate, yet some have both sweetness and flavour, though the average consumer finds it difficult to know, or remember, which ones. However, that intensity of flavour does not seem yet to appear in the larger summer tomatoes which are grown commercially.

The best tomatoes for sheer flavour grow in Mediterranean climates – such as those of southern Europe, California, Mexico and so on – in poor soil, often withstanding drought. In fact, tomatoes obviously like to struggle for their survival and their nourishment, and this alone gives them a concentration of flavour, even if the fruits are misshapen and small.

NUTRITION

The highest vitamin content is in the jelly which surrounds the seeds, so deseeding tomatoes reduces their food value. They are high in fibre, potassium and folic acid. A good fruit, then, to eat throughout pregnancy.

VARIETIES

The hundreds of varieties can best be classed according to their size and shape. Some grow outdoors where summers are warm enough; in temperate climates many are grown under glass.

ROUND or SALAD TOMATOES: the standard commercial crop, available all year in some form, but best when sun-ripened in summer. Relatively juicy, with high seed content. Moneymaker, Ailsa Craig and Alicante are reliable; Harbinger is an old variety that some people consider has more flavour.

CHERRY TOMATOES: small-scale versions of the above, often with sweet, well-flavoured fruit. Some plants are small enough for container-growing. Gardener's Delight is a favourite, as the name implies, both sweet and juicy. Sweet 100 is prolific and delicious. Phyra is very small, like a wild tomato. Pixie and Tiny Tim are particularly compact plants.

PLUM TOMATOES: relatively fleshy, with fewer seeds and less juice. The Italian tomato that is generally canned. Needs good sun to ripen well. San Marzano is the variety to grow.

LARGE or BEEFSTEAK TOMATOES: outsize fleshy fruit with relatively few seeds; most need hot sun to ripen to full flavour. Oxheart Giant has huge pink fruit shaped like a heart. Marmande is my own favourite, irregularly shaped but with the finest flavour.

YELLOW TOMATOES: colour variations on round types include Golden Sunrise, Golden Boy and Yellow Perfection. Tigerella is striped in red and yellow.

CHOOSING AND STORING

The flavour is finest in fruit plucked from the plant after the sun has been on it, but most of us now buy our tomatoes from the supermarket. A sign of freshness is a leafy-looking green calyx, not one that is withered and spidery. Keep tomatoes in a cool, dark place – the salad compartment of the refrigerator is fine – but take them out to reach room temperature before eating. Cherry tomatoes come in polythene boxes and keep well in these for a week or more in the refrigerator.

In the short summer season there are often gluts. Fortunately much can be made of the less-than-perfect tomato. Rather overripe ones are fine for sauces and soups (see Cooking, below), which incidentally freeze well. At the end of the season gardeners often strip plants and put any unripe tomatoes in a drawer or cupboard where they ripen slowly during the following weeks. Unripe green tomatoes also make excellent chutney and soup.

Among commercial products on the market canned plum tomatoes are worth keeping in the store cupboard: out of season their sun-ripened flavour is preferable for sauces to insipid fresh fruit.

TOMATO JUICE: This will keep well as long as it is unopened. Once the pack or can is opened, use within days.

TOMATO PURÉE: Canned tomato purée must be used up once it is opened, or it can be decanted into a jar, covered with olive oil and kept in a refrigerator. But even with this precaution it will not keep much longer than a week. It is best to buy tomato purée in a tube. Store it in the refrigerator once opened, and it will keep for months.

DRIED TOMATOES: Commercial sun-dried tomatoes can be bought loose or packed in oil in jars. Store loose ones in airtight jars, and reconstitute by covering with boiling water and soaking for about 15 minutes. Keep the others in their oil, and when the tomatoes are all eaten use the oil for sauce, vinaigrette or cooking.

DRYING TOMATOES: In the absence of warm sun or a fan-assisted fruit dryer, make your own dried tomatoes using a very gentle oven (the tomatoes do not cook, but simply dry out and contract as all the moisture is driven off). Cut the tomatoes in half (scrape away some seeds if they are very moist), sprinkle a little salt, sugar and oregano on each one, lay them skin side down on a baking tray and place in a very low oven for a few hours. If you have a solid-fuel stove, you can place them in the bottom of the oven with it merely turned on. These tomatoes are best stored in olive oil or they will go mouldy rapidly. The oil will also develop a lovely flavour and it is very good in salad dressings.

PREPARING

Many recipes tell you to remove tomato skins and seeds before cooking. For one thing, both are indigestible. Indigestible means what it says in this case (it is not an euphemism for producing flatulence): the skins and seeds go right through the gut unchanged and are excreted. There is nothing harmful about this for you or the tomato. However, both the skins and the seeds look unsightly in many recipes or spoil the texture. I never bother to 'deseed' the tomatoes as requested, but for most soups and sauces it is best to sieve them. Skins and seeds are more obtrusive in some varieties and in some individual fruits than in others. Use your judgement about whether or not to remove them.

To remove skins, plunge the tomatoes into boiling water, or pour boiling water over them, and leave for a minute or so. Puncture the tomato; the skin will split, and is easily peeled away. This would be necessary, for example, for a tomato and mozzarella salad where I would want the tomatoes to soak up the flavours of basil and oil, which only a skinless tomato will do.

For a well-flavoured tomato soup, I would include not only the green calyx but if possible some green leaf too. This is only possible if you grow your own tomatoes. It is from the greenery that the powerful tomato aroma emanates. Yes, I know I have said it is toxic, but in this tiny amount I have never known any vague disquiet after eating the soup and it does make a difference to the final flavour.

Green tomatoes need not be skinned; they can be cut up for chutney as they are and the soup below uses the entire thing.

COOKING

For soups and sauces, throw chopped tomatoes into a dry pan, cover, place over a very low heat and leave to cook in their own juice. Done within 10–15 minutes. Then liquidize and put through a sieve. The debris is thrown away. This is a real home-made tomato purée and it tastes fantastic. To the tomatoes, before cooking, you can add cloves of garlic, basil, a dash of olive oil, but nothing is really necessary, for the tomatoes *au naturel* are excellent. About 450 g / 1 lb of tomatoes will make roughly 300 ml / ½ pt of sauce, so in order to make this soup for six people you should expect to need 2.7 kg / 6 lb of tomatoes. Oddly enough, this is not the case: the more tomatoes you use, the greater the ratio of juice to tomato. However, it also depends on how ripe the tomato is. It is best to use tomatoes which are over the hill and being sold off cheaply, for these are by far the juiciest. Using these I have been able to make enough soup for six people out of 1.8 k / 4 lb of tomatoes.

For a sauce, make your juice in the above manner, then reduce over a high heat to intensify the flavour. Add at the start of cooking a little alcohol – wine or spirits. One of the best tomato sauces I ever made had a slug of malt whisky added at the end of cooking. It diluted the sauce, yes, but oh, what flavour.

CANNED TOMATOES: In the winter making a good, well-flavoured sauce from canned tomatoes would seem even easier, but canned tomatoes need to be lifted from their own banality, so I would add various other flavourings. Firstly, sweat some garlic and onion in olive oil, then add a pinch of oregano or marjoram, 2 diced boiled carrots, seasoning, a glass of red wine and then the can of tomatoes. Cook for 5 minutes, then season and blend to a smooth sauce.

TOMATO PUDDING

Based on summer pudding, this recipe was devised by Jennifer Patterson – the eccentric and boisterous *Spectater* cook. Alas, since I have lost her book, I cannot say I am indebted to her for this particular version. However, the idea is to use a 1.5 litre / 3 pt pudding basin – these amounts will fill it about halfway.

1 small loaf of day-old bread, sliced
1.1 kg / 2½ lb tomatoes
1 garlic clove, crushed
few basil leaves, chopped, plus more to garnish
1 tsp caster sugar
1 tsp sea salt
1 sachet (25 g / ¾ oz) of aspic jelly, or 2 tsp gelatine

First line the pudding basin with slices of white bread, after cutting off the crusts. Fit the pieces neatly together so that the bottom and sides are covered, with the bread reaching about halfway or a little more up the basin. Cut out a piece to fit the top. Reserve.

Peel the tomatoes by pouring boiling water over them and leave for 2 minutes. Remove the skins, then chop the flesh coarsely. (Do this in a soup plate, or a similar utensil, so as not to lose any juice.)

Place the tomatoes in a pan with the garlic, basil, sugar and salt. Bring to the boil, turn the heat off, immediately melt the aspic jelly or gelatine in the hot tomato.

Pour the mixture into the bread-lined pudding basin. Make sure that as much as possible of the diced solid tomato is packed in: you need the pudding to be as firm as it can be. Place the bread 'lid' over. You may have plenty of juice left over; this doesn't matter. Pour it into a basin to set and serve with the pudding. Cover the bread lid with a plate of a size which will press down on to the pudding and place a weight on the plate. Leave for a day in a cool place.

Unmould and decorate with the spare jellied tomato sauce chopped up around the base and a few basil leaves scattered over the pudding itself.

RIGHT *Red and yellow tomato chutneys (see overleaf) served with grilled aubergine slices*

TOMATO MOULDS

for 6

handful of basil leaves
900 g / 2 lb very ripe tomatoes
2 garlic cloves
1 tsp sea salt
1 tsp caster sugar
1 sachet (25 g / ¾ oz) of aspic jelly,
or 2 tsp powdered gelatine
2 glasses of dry white wine

Reserve 8 basil leaves, but throw the rest into a saucepan with the tomatoes, the garlic, salt and sugar, place a tightly fitting lid on and leave over a low heat for 10 – 15 minutes. Leave to cool, then liquidize. Pour through a sieve, throw away the debris and reserve the sauce.

Melt the aspic jelly or gelatine in the white wine and stir this into the tomato sauce.

Place a basil leaf at the bottom of each ramekin and fill with tomato sauce. Refrigerate for 6 hours. Unmould on to individual plates 5 minutes before serving by dipping the ramekins in hot water and easing the contents out with a knife. A sauce of yellow peppers goes well with this, but some rocket leaves I regard as essential. Alternatives you might like to try are sour cream and green peppercorns, or a sauce of fresh green peas or leeks.

GREEN TOMATO CHUTNEY

1 kg / 2 lb 4 oz green tomatoes
2 heads of garlic, peeled and sliced
600 ml / 1 pt apple cider vinegar
2 green chillies, chopped
50 g / 2 oz root ginger, peeled and grated
2 tbsp brown sugar
1 tsp sea salt

Chop the green tomatoes coarsely and throw them into a preserving pan with the rest of the prepared ingredients. Bring to the boil and simmer, stirring the mixture now and again, until the tomato chutney begins to thicken a little. Taste: it may need a little more sugar or salt. The chillies and ginger make this a fairly spicy chutney.

Leave to cool, then bottle the chutney in sterilized jars. It will keep for months if need be. Or, it can be eaten at once.

Note: The same chutney can be made with red tomatoes, but because the fruit is ripe, you will not need the same amount of vinegar. So cut the amount by half, and use caster sugar instead of brown. Also the red chutney will disintegrate into something more like a rough purée.

GREEN TOMATO SOUP

This is an excellent way – other than making chutney – of using up a glut of green tomatoes.

2 tbsp olive oil
900 g / 2 lb green tomatoes
2 garlic cloves, chopped
2 tbsp caster sugar
1 tsp sea salt
300 ml / ½ pt buttermilk or smetana, or 3 tbsp fromage
frais, or 150 ml / ¼ pt sour cream
basil leaves for garnish

Heat the oil in a saucepan. Chop the green tomatoes and throw them into the pan with the garlic, sugar and salt. Bring to the boil and simmer for 1 minute, no longer.

Let the soup cool, then liquidize in a strong food mixer, adding at the last the buttermilk, smetana, fromage frais or sour cream.

To serve, float some chopped basil leaves on the surface of each bowlful.

ABOVE *Green, yellow and red tomatoes*

AUBERGINE

Solanum melongena

Though the aubergine or eggplant is an eminent member of the potato family, it was not discovered by Columbus in the New World. It comes from India, where it is believed to have been grown and eaten for more than four thousand years. It took some time to leave South-east Asia, the ancient Mediterranean world was ignorant of its charms, and it is not until the thirteenth century that the great German scholar Albertus Magnus mentions it.

It was the Moors of Andalusia who brought the plant to southern Europe from the Middle East. The derivation of the name 'aubergine' – from the original Sanskrit through Arabic to Catalan and finally French – reflects this route. The vegetable was certainly cultivated in Sicily and southern Italy, where Moorish influence was strong, but gained only a limited acceptance farther north. It appeared upon the menu of a banquet given by Pope Pius V in 1570, but aubergines in general were considered unhealthy. Anna del Conte tells us that a fourteenth-century writer regarded them as 'causing males to swerve from decent behaviour'.

One traditional dish remains part of the Italian repertoire – *melanzane fritte*, where the aubergine is sliced, coated in batter and deep-fried in olive oil. But it took almost to the present time for the aubergine to be generally accepted in Italy and elsewhere. Jean de la Quintinie, gardener to Louis XIV, first grew aubergines in France; Thomas Jefferson, who delighted in new foods, first grew them in America, but they remained a curiosity and did not really catch on in either country. It is only with the widening of food tastes all around the Western world in the last generation or so that aubergines have gained any of the popularity that they enjoy in Asia and the Middle East, where they are a central part of the cuisine. One reason for this could be that of all vegetables the aubergine most radically changes its flavour, texture and appearance when cooked and, secondly, that its flesh soaks up and fuses completely with added flavours. This is its prime quality and function in a whole range of Indian cooking. So wherever you find a cuisine which uses many and varied spices, you tend to find the aubergine.

In India and America you will often find white or creamy aubergines roughly the size and shape of an egg, so it is not difficult to see how the name 'eggplant' came into being. There are, in fact, dozens of varieties from the dark purple marrow shape we know well to oval and round ones in lilac shades, and others striped white and green. In fact, the common practice of Indian restaurants in the West of adding green peas to curries stems from their chefs' attempts to reproduce the look of a curry which contains little whole green aubergines the size of quails' eggs, or even smaller.

NUTRITION

Aubergines are rich in vitamins B1, 2 and 3 and vitamin C. They also have small amounts of iron and calcium. The skin, as always, is richer in nutrients than the flesh.

VARIETIES

Aubergines need warmth to grow: glass protection is usually called for in temperate climates, except in particularly sheltered sites and warm seasons.

There is relatively little difference in flavour between the varieties, and for once the size of the fruit has little to do with flavour and texture. Choice is a matter of personal preference – grow varieties that will do well in your conditions and that suit the dishes you plan to cook.

Among the purple-skinned types of aubergine, Onita is early and productive. Short Tom is best picked young, as the name implies. Black Prince is a hybrid from Japan. Black Enorma can grow to 675 g/ 1½ lb in weight. Easter Egg is small and creamy in colour, about the size of a hen's egg.

STORING

Aubergines will keep in the salad drawer for two weeks or more if they are fresh. Inspect the calyx for colour and perkiness; if dry and brown, the aubergine has already been kept too long. Aubergines should be firm to the touch and glossy in colour; when old they get soft and a little limp and lose their sheen.

PREPARING

Aubergines must be cooked: they are unpalatable raw. They can be prepared in different ways depending on the finished dish. Some recipes counsel putting cut aubergines in acidulated water to prevent discolouring, but since the flesh anyway changes colour when cooked I think this is unnecessary. Many older recipes also advise you to salt the flesh and leave it to drain to remove bitterness. This is no longer a problem with most cultivars available in the West (I steam aubergines for just a few minutes and there is not the slightest trace of bitterness).

The real value of salting is to counteract the tendency of aubergines to take up great quantities of oil. This is due, so Harold McGee tells us, 'to the very spongy texture of its tissue: a high proportion of its volume consists of intercellular air pockets'. He goes on to say that there comes a point when the heat of the pan and the concentration of oil induces a 'collapse of structure' and the aubergine slices expel their oil. This is frankly a sorry and unpalatable mess. We do not want aubergine slices either full of oil or after expelling their oil, so the secret is to get rid of those 'intercellular air pockets'. One way of doing this is by slicing the aubergine, salting the slices, leaving them out in the air for an hour or more, then rinsing the salt away and patting them dry. The slices can then be floured, dipped in egg and breadcrumbs or dipped in batter and fried. Or they can be fried on both sides without such additions. Fry them in olive oil over a high heat until they are brown and crisp – a few minutes. For a purée the whole aubergine can be pricked and then baked in an oven at 190°C/ 375°F/gas5 for 30 minutes, left to cool, then sliced, the interior flesh taken out and the skin thrown away. Alternatively, it can be boiled for 15 minutes or steamed for 20 minutes. The quickest method I know

is to cut the aubergine in slices and to steam them for 10 minutes.

If the aubergine has been boiled, baked or steamed, the air pockets will all have vanished, so the flesh will not soak up enormous amounts of oil. However, adding oil to the aubergine flesh is a little like making mayonnaise: the flesh still seems to soak up an astonishing amount. Watch this tendency. It is impossible to give quantities, as aubergines differ in size and the flesh is of various consistencies. So taste the results continually.

Whole aubergines can be halved, salted and left for an hour before being cooked. Or they can be sliced lengthwise quite thickly, salted, left, washed, drained and then barbecued, sandwiched between a turning grid which chars the vegetable attractively.

COOKING

Aubergines have many uses in a great variety of dishes. Claudia Roden lists 33 recipes in her classic *Book of Middle Eastern Food*, from aubergine meatballs to stuffed aubergines and sweet and sour salads. Firstly, they turn up as appetizers – purées made with yogurt, olive oil and lemon juice, served with olives, sweet peppers and pitta bread.

Aubergines have entered legend. Not only do we have the famous *Imam Bayildi* – 'the Imam fainted' – but Claudia Roden gives us another, Sultan's Delight. Both recipes are not particularly spectacular in themselves. The truth is that the aubergine puréed, with the addition of other flavourings or vegetables, is remarkably seductive. We are inclined to forget how astonishingly delicious this vegetable can be. So here are a few suggestions for various aubergine dishes where it becomes puréed, which can be eaten hot or cold.

For the first, steam the sliced aubergines for 10 minutes, then heat a few tablespoons of olive oil with 2 crushed garlic cloves, several skinned tomatoes, 2 sliced courgettes and the steamed slices of aubergine. Cook for 10 minutes, then add a tablespoon of soya sauce and some crushed basil leaves.

Alternatively, steam the sliced aubergines for 10 minutes, then heat a few tablespoons of olive oil with 2 cloves of garlic and ½ teaspoon each of coriander, cumin and mustard seeds, 2 chopped onions and 1 teaspoon turmeric. Throw in the aubergine and cook for another 10 minutes.

In both these cases the aubergine will disintegrate into a purée, leaving the strips of purple skin, but absorbing the flavours. Both can be eaten as they are, hot, warm or cool, or as a sauce with pasta.

Eggs can be broken and dropped into small indentations in both purées and cooked for a few moments.

A third option is to bake or boil 2 aubergines. When cooked, scrape all the flesh into a blender, add juice of 1 lemon, 1 crushed clove of garlic, 2 spoonfuls of yogurt and 4 tablespoons of olive oil. Blend to a thick cream. This is a basic purée to be used as an appetizer. But additions can be made to it. Chopped red or green chillies will make it fiery; chopped tomatoes, a spoonful of tomato purée and a teaspoon of paprika will give the aubergine a tomato subtext, while a dash of sesame oil and tahini plus roasted sesame seeds give you a nutty and highly delicious hummus-like purée which is characteristic of the Middle East.

CAPONATA

1 medium to large aubergine
3 tbsp olive oil
1 onion, thinly sliced
225 g / 8 oz tinned chopped tomatoes
4 tbsp red wine vinegar
1 tbsp caster sugar
2 tbsp capers
10 stoned black olives
10 stoned green olives
sea salt and freshly ground black pepper

Slice the aubergine lengthwise, then dice these slices into cubes; salt the cubes and leave in a colander for 1 hour. Wash under cold running water and pat dry with a clean towel.

Heat the olive oil in a pan, throw in the onion and aubergine and fry until the aubergine is crisp and brown.

While this is happening, in another saucepan heat the chopped tomatoes, wine vinegar and sugar. Let the sauce simmer for a few minutes, then add the capers, olives and seasoning.

Turn out the aubergine and onion mixture into a serving dish and pour over the tomato sauce. *Do not* cook the aubergine in the sauce, or the aubergine cubes will disintegrate.

Note: In the original Sicilian recipe, celery or artichokes are cooked with the aubergine and everything is eaten in the sweet-and-sour sauce. Chocolate is also sometimes added to the sauce with toasted almonds. As there is no celery worth eating in the summer months when the best aubergines are available, I have not added it. But you can add any other vegetable you wish to the cubed aubergine and onion as long as they remain separate and do not disintegrate into a mush when cooked in the oil. I suggest baby courgettes, leeks, carrots and turnips.

AUBERGINE SESAME

3 medium aubergines
1 garlic clove, crushed
juice of ½ lemon
1 tbsp toasted sesame oil
6 tbsp olive oil
sea salt and freshly ground black pepper
4 tbsp sesame seeds

Prick the aubergines and place them in an oven preheated to 180°C/350°F/gas5. Cook for 40–45 minutes, then feel them: if soft, they are done.

Leave to cool, then scrape all the flesh out and place in a blender. Add the garlic, lemon juice and sesame oil. Blend, then with the machine on low add the olive oil slowly, on the same principle as when making mayonnaise, but at a slightly faster rate. When the oil is absorbed and the aubergine has become creamy, season with salt and pepper.

In a deep frying pan heat the sesame seeds until they begin to pop and turn golden brown. When these have cooled, add 3 tablespoons to the aubergine and mix them in.

Turn out into a serving bowl and sprinkle the last tablespoon of seeds over the top. Eat with hot pitta bread or crudités.

SWEET PEPPER

Capsicum annuum Grossum Group

This is a shrubby annual plant which bears fruit within a couple of months. The glossy fruits are roughly bell-shaped (hence bell peppers) and hollow: their botanical name derives from the Latin *capsa* meaning case or box. At first green, the fruit turns yellow and then red if left to ripen on the bush. Compare this description with that of another kind of pepper. The peppercorns we grind up and use to season dishes on the table come from a vine (*Piper nigrum*) that can grow to 9 metres / 30 ft. This begins to fruit after three years and will continue for 20 more, and bears its tiny green berries on long spray-like stems. One would think no two plants could be more dissimilar. However, when Columbus anchored off the Caribbean island he named Hispaniola (now Haiti and the Dominican Republic), he was convinced he had found the Spice Islands off the coast of India. Therefore, any vegetable that was fiery and used as a seasoning must be a pepper. When the ships returned to Europe they carried with them 'peppers of many kinds and colours...more pungent than that from Caucasus', as Peter Martyr wrote when he observed the triumphal arrival of Columbus at the Spanish Court in April 1493. Peter Martyr was not convinced these new vegetables were really pepper, though he adds 'it is just as much esteemed'.

On the second Columbian voyage it was noted that the main staple in the natives' diet was a bread made with sweet potato eaten with a hot and peppery seasoning which was used extensively with fish and any small birds the natives caught. The same seasoning they found in the highlands of Mexico at the court of Montezuma. When Cortes and his Spanish army reached this civilization they were impressed by the great range of dishes, including turkey served with a sauce of tomato, chilli and ground marrow seeds, and fish with sauces of yellow and red chilli.

By then the Spaniards had christened the new vegetable (whatever the size, colour, shape or degree of fieriness) *pimiento*, after their word for black peppers – *pimienta*. If that is not confusing enough, another berry, which does resemble the peppercorn – allspice – was also named *pimiento*.

Thus the whole of the large capsicum family is now called peppers, though the small fiery one which started the misunderstanding took on the Aztec name, chilli. The Spanish brought that word back to Europe and very quickly it spread to the rest of the world.

The Spanish found that the sweet peppers as well as the chillies grew well in southern Spain and in their new possessions in Sicily and southern Italy. Very soon sweet peppers spread around the Mediterranean. Their appearance was keenly appreciated: 'a crimson and scarlet mixt: the fruit about three inches long and shines more than the best pollisht corall', says Lignon in his *History of Barbadoes, 1647–53*.

Today, familiar with the vivid colours – red, orange, yellow, purple and black – that peppers display, we might imagine that some are recent and have been selectively bred. However, Peter Martyr's description belies such a thought. The Columbian pepper harvest which reached Spain included peppers of all colours, each with a different name. There was even a white pepper – both within and without. Another was violet without and white within.

No other vegetable among the Columbian discoveries was accepted so readily as the sweet peppers and their fiery nephew, the chilli. (Some chillies belong to the *Capsicum annuum* species, others are kinds of *C. frutescens* – see page 206.) They became absorbed into southern European and Asian cuisine within the next half century, so much so that many of the best pepper dishes are Mediterranean in origin.

Paprika is made from dried sweet peppers. The type used is generally the European sweet peppers, which are large and mild. Hungary is the main producer of paprika; the more fiery the colour, the greater the quality. In Spain they also grind sweet peppers and there they call it *pimentón*. This is virtually the same as paprika and, again, the more vivid the colour, the better it is.

NUTRITION

Sweet peppers are high in vitamins B1, B2 and C and are also rich in fibre, potassium and folic acid. They are an excellent salad food if eaten raw.

CHOOSING AND STORING

The outside should feel hard and crisp; patches of softness or wrinkling show that the fruit is no longer fresh. The colour should be glossy and vivid. If peppers are in prime condition when bought, they will keep for a few days. But they quickly fade and soften, so use them within three days.

VARIETIES

It is well worth growing your own peppers, outdoors in warmer areas, or in a greenhouse in most temperate zones. Peppers eaten just after picking have a quality and flavour that is sensational, both sweeter and more pungent and peppery than any bought from a store. The different colours do not have markedly different tastes.

NEW ACE: Green bell shape: a vigorous early variety.
LONG RED MACARONI: Deep red pods and a mild sweet flavour.
LONG YELLOW: Exactly as the name suggests – sweet and mild.
CANAPE: A green bell shape which can be grown outdoors in summer.

PREPARING AND COOKING

For all recipes peppers need to be cored and deseeded. Simply cut off the top with the stalk (for stuffed peppers you will need this top as the lid), then with a sharp knife gouge out the core with its seeds. Most peppers will also need the pithy lining cut away too, which is easily done if the pepper is now sliced in two.

For certain dishes peppers need to be peeled. The outer skin, though thin, is tart and slightly bitter. This gives a pleasant stimulus to salads, but once cooked the outer skin becomes acrid and mars the sweetness of the pepper.

The easiest method of peeling is to use a potato peeler, preferably the swivel type where no pressure is needed. However, some dishes call for a smoky flavour in the peppers, which you can best achieve by blistering the skin over hot charcoal, then leaving them to cool before scraping off the skin. This can be tedious. You can also hold peppers over a gas flame with a pair of tongs, which will similarly give you a slight smoky flavour. Popping the charred peppers in a plastic bag to sweat for a few minutes before peeling helps the skins come off more easily.

But how to decide for any particular recipe whether the peppers should be peeled or not? Or by what method? This subtle problem can only be decided by trial and error. Try a favourite recipe with the peppers peeled, charcoal-grilled, or left completely raw. It is a matter of personal taste.

Peppers have been used for centuries in Provençal cooking. About twenty years ago there was a great vogue for one particular dish – ratatouille, a mixture of peppers, aubergines, onion, courgettes and tomatoes stewed softly in olive oil. Unfortunately, this dish used to be overcooked, so that an unappetizing oily brown sludge would appear. Be sure when adding peppers to other vegetables to stop the cooking process before the individual vegetables disintegrate. (See my recipe below.)

Peppers will cook, when chopped and sautéed over a low heat in oil, in about 10 minutes. If stuffed and baked in the oven they will take about 30 minutes. I confess I am not an enthusiast of stuffed peppers (see Claudia Roden's book for the classic recipes), for I dislike the flavour of peppers when they have been cooked that long. Also, long baking makes the outside indigestible and the interior too soft to be enjoyable. If you want peppers and rice, I suggest you cook them separately and quickly, within 10 minutes.

However, the first pepper recipe I give here is an old favourite, one of the very first dishes I learnt to cook in the 1950s. It makes an excellent meal for summer lunch with a salad.

PIPÉRADE

3 tbsp olive oil
675 g / 1½ lb mixed peppers, cored, deseeded and sliced
2 large onions, sliced
pinch of oregano
675 g / 1½ lb tomatoes, peeled, or 225 g / 8 oz tinned chopped tomatoes
4 eggs, beaten
sea salt and freshly ground black pepper

Heat the oil in a frying pan and throw in the peppers and onions with a pinch of oregano. Cook, stirring occasionally, for 5 to 8 minutes, then add the tomatoes, turn up the heat and cook for a further 3 minutes. (If the tomatoes are fresh this is all the cooking they will need. The mixture should be sloppy. If the tomatoes are tinned there may be too much liquid, so raise the heat for a moment or two so that most of this evaporates.)

Now pour in the eggs and mix thoroughly as you would for scrambled eggs; the mixture should just set. Take from the heat immediately and serve with good crusty bread.

RATATOUILLE

3–4 tbsp olive oil
3 mixed peppers, cored, deseeded and sliced
1 onion, sliced
3 small courgettes, cut into chunks
1 small aubergine, diced, salted and left for 1 hour, then rinsed and dried
sea salt and freshly ground black pepper

Heat the oil and cook the vegetables at the same time, stirring occasionally. Cook for 12–15 minutes, no longer, so that the vegetables are softened and browned, but still in their individual pieces. Add seasoning at the end, because salting makes the juices run. Eat warm, with good crusty bread.

GRILLED PEPPERS AND ANCHOVY

1 large red and 1 large yellow pepper
1 tin of anchovies
1 small red onion, cut into rings
5 black olives, stoned
1 tbsp olive oil
1 tbsp capers (optional)
juice of ½ lemon
freshly ground black pepper

Peel the peppers with a potato peeler then cut them in half and scoop out the seeds and core. Slice them in strips and lay on a large round oven dish, keeping red and yellow colours separate in four quarters.

Lay the anchovies in two vertical and two horizontal lines. Lay a few onion rings in each square, with an olive half in the centre. Dribble a little oil over the lot and sprinkle a few capers over the top.

Place under a preheated very hot grill for about 2 minutes. The peppers and onions should blacken a little, but not necessarily lose their rawness.

Give a good squeeze of lemon juice over the dish, plus a few turns of the pepper mill, and serve.

THREE COLOURED PEPPER PURÉES

Peppers come in the most vibrant colours – red, orange and yellow. There are also black peppers; however, if you cook with them, the black turns green. You can, of course make a purée from green peppers but I much prefer to use the red, orange and yellow – adjacent colours on the spectrum – for a more stunning effect. The purées are simple to make. Follow the method given here for the red peppers for the other colours.

2 garlic cloves, peeled
2 red peppers, cored, seeded and sliced
2 tbsp good olive oil
pinch of sea salt

Heat the olive oil in a saucepan, throw in the peppers, garlic and salt. Let everything sauté very gently beneath a close-fitting lid so that the peppers steam in their own juices. Cook for 12–15 minutes, then let them cool.

Blend in the liquidizer to a thick purée. The oil forms an emulsion with the peppers.

Do the same with the other two colours. Keep in separate jugs. Use as a coulis around moulds or for a first course where the food is in the middle of the plate. Pour into circles or triangles of colour. For the last, use a piece of card to keep the colours separate while you pour them on the plate. They are thick enough to stay where you put them.

RIGHT *A parsnip mould (page 232) served with the three coloured pepper purées*

CHILLI PEPPER

Capsicum annum Longum Group and *C. frutescens*

After salt, the chilli pepper is now the most popular seasoning in the world. There are dozens of different chillies which come in various shapes from small bells and round puffs to long, thin, tapering cones, all having varying degrees of fieriness. The chilli pepper was an instant and worldwide success almost as soon as it was discovered, such was the European desire for a spice akin to pepper. Because of the Portuguese trading posts around the coast of Africa to the Indian Ocean and beyond, fifty years after the chilli had been discovered it was known and sold in barter from Africa to Nagasaki. From then on the chilli began to flavour cuisines from Gambia to Bangkok. In every region where it could be grown it was used – dried, ground and whole – appearing in diverse dishes giving a scorching zest to the staple diet.

The heat is due to a substance called capsaicin, which is in the pith rather than in the seeds, which many cooks discard. To get the maximum heat from chillies, ensure that the pith is cooked; to reduce the hotness, take care to eliminate it. Nor are red chillies hotter than green ones. Red simply means ripeness and a greater sugar content.

New scientific research by Dr John Prescott of the Australian national research organization, CSIRO, has discovered that capsaicin acts as a flavour enhancer to other foods. It is thought that capsaicin triggers the release of endorphins, the body's natural painkillers, which creates a sense of general well-being and thus makes the pleasure of eating more intense. That is not all that capsaicin is thought to do (see the later notes on nutrition).

It is difficult to tell how much capsaicin is in a pepper – size is no guide, as more of the substance is likely to be in the small bell-shaped peppers than in the larger thinner ones. If the plant has to struggle for survival, the amount of capsaicin will increase. So poor soil, hot sunshine and inadequate watering will create a chilli small but powerfully hot, while a plant grown in my own polythene tunnel in temperate Suffolk in rich soil and watered liberally will only be moderately hot. Capsaicin reaches its peak in the matured ripened chilli, but here the heat is offset a little by the sweetness.

The beneficial effects of capsaicin might help to explain the popularity of Tabasco sauce, which contains chilli peppers, salt and vinegar. The recipe for Tabasco was devised by Edmund McIlhenny for his own use in the 1850s. He was a New Orleans banker who had been given a handful of dried peppers by a friend on his return in 1848 from the war in Mexico. Edmund planted the seeds in his father-in-law's plantation, Avery Island, in southern Louisiana. The plants flourished and there he made his sauce and gave bottles away to a few friends. Fortuitously, the island had a salt mine, so a pepper sauce was a perfect complement. The salt was used a few years later, in the American Civil War, to provision the Confederate troops, impelling the Yankees to invade the island and take possession of the mine for their own use.

After the war, with the Southern states defeated and the economy in ruins, Edmund saw no future in banking. But the chilli pepper had taken over Avery Island and by 1869, 350 bottles of Tabasco Pepper Sauce had been dispatched to selected wholesalers. The product never looked back.

Tabasco is an Indian word meaning 'land where the soil is humid' and certainly Louisiana – with its bayous and swamps, its trees hung with Spanish moss like the swagging at Miss Haversham's wedding feast – is a place where in the summer you steam lightly. The seeds are planted and harvested in August, the pickers all carry a *baton rouge* the exact colour of the ripe chilli, so there can be no mistake in plucking ones still unripe. The peppers are then mashed and placed in oak barrels with salt and left to mature for three years. A wooden top with holes is covered with a layer of salt. This allows the gas to escape when the peppers are fermenting, but the salt insulates the contents from any bacteria. The factory is thick with fumes, and the air sharp and acrid. The pepper stings and catches the throat, seeming almost to scrape the lining of the windpipe. On a recent visit many food writers coughed constantly, even though they had covered their mouths and noses with scarves and cardigans. But the workers we spoke to had adjusted to this peppery intake and swore they never had coughs or colds, or any form of sinusitis or hay fever.

After three years' fermentation the mashed chillies are tipped into hundred-gallon vats and white vinegar twice the strength of table vinegar is added. The contents of the vat are kept moving for 28 days and finally the juice is strained and poured off.

NUTRITION

For such a small, fiery fruit, they are surprisingly high in vitamins A and C. Capsaicin has beneficial effects on our digestive system, its ingestion stimulating the saliva, gastric juice and the peristaltic movements of the gut. It is thus good for constipation, for the capsaicin irritates the intestinal lining and increases its movement, cutting down the time that the meal takes to pass through. It is also good for the circulation and is a healing element for stomach and intestinal ulcers. It is also an antiseptic, actually used upon sticking plasters. A little chilli used discreetly every day – like garlic – in at least one dish would undoubtedly have a beneficial effect on anyone's diet.

STORING

Chillies may be bought fresh, dried, canned and pickled. They are useful to have in the store cupboard in all these guises. Fresh green chillies will, like sweet peppers, get limp and wrinkled after about a week and must be used or thrown away. They keep for a little longer if stored in the salad drawer of the refrigerator. Red chillies, on the other hand, will, if not used within a week or two, dry out and can be kept for many months, as long as they have dried out hanging up within an airflow.

Dried red chillies can be bought and kept within a clean jar and they will keep forever. I have a Victorian spice tin which belonged to my grandmother, still with some tiny red chillies in it, possibly used by her about 1900 – who knows? I keep them for sentiment as a curio. I don't advise anyone to cook with such ancient spices. But I know that some dried red chillies (as I tend to buy them in bulk) I must have had for five years or so.

Dried chillies can also be purchased flaked or ground to a powder, or in the form of Tabasco and a list of other hot sauces. Cayenne pepper was thus named after the capital of French Guiana, which lay on the Cayenne River. However, chillies are not grown in this area any more, and cayenne pepper can now be made from chillies grown anywhere in the world, from Louisiana to India and the Far East. There is very little difference between cayenne pepper and chilli powder, though commercial chilli powders may have colourings and other spices added to the ground chillies.

Both cayenne pepper and chilli powder will lose their pungency far quicker than a dried red chilli. Buy these in small amounts, keep them in a cool, dry and dark place and do not expect them to taste as fiery after a year or two as they once did.

VARIETIES

Away from semi-tropical regions, chillies must be grown under glass. They are, however, easy to grow once they have germinated and are prolific in fruiting. One small plant a few feet high could give you over a hundred chillies. None need be wasted, for they can be threaded and hung to dry.

CHILLI SERRANO is a long red chilli that is very hot.

EARLY JALAPEÑO starts dark green and turns to red. This is one of the most popular chillies in America, where they are picked and served as a relish. In Mexico the jalapeño is smoked and becomes a '*chipotle*'.

ETHIOPIAN is a long thin pod that gets hotter and redder as it ripens.

POBLANO is a small dark green chilli which is quite mild and can be roasted or grilled whole. In Spain they serve plates of them in the tapas bars. They are delicious, but astonishingly inconsistent: there will be one among twenty which is scorching hot, while the rest have been sweet and spicy. It is like playing Russian roulette with one's palate, but well worth it.

CASCABEL can be bought dried. Small, plum-shaped, with a nutty flavour, also roasted whole, but used too to make *salsa de chile cascabel*.

HABANERO a small green, yellow or red chilli from Mexico that belongs to a separate species (*Capsicum sinense*). It is the hottest of all and is commonly called Scotch Bonnet.

PREPARING

Take care when handling chillies as capsaicin is a skin irritant, especially painful if any gets into scratches or cuts – or worst of all, and most common, transferred to the eyes by rubbing in the eyelids. Infinitesimal amounts nestle beneath the fingernails and can irritate delicate skin even some hours after handling. It must be stressed that capsaicin, though beneficial, has this aggressive aspect; it is a general irritant and is used as an active component in anti-mugging services. When preparing chillies, to ensure utmost safety, either use gloves or handle the chilli with a knife and fork (my method) and wash the implements afterwards. The chilli can be quite easily dissected using a sharp knife and sliced, diced small or (with pith and seeds removed) left in a chunk.

COOKING

Chillies are often diced small and used with other spices, sweated in oil at the start of creating a dish, so that the fieriness will permeate through. However, with some Asian and Indian recipes, one whole chilli can be added whole and left to permeate the dish and give another aspect of heat. I am personally fond of this method; it is subtle and the heat has a quite different quality. There are other recipes where the chilli is added last. Chopped and diced small, it is fried with a mixture of spices and then stirred in at the last moment. Several lentil dishes use this method, where the bland pulses suddenly acquire a small pool of fire. Highly satisfactory, too. Those nervous of the heat can avoid the spicy bits if they wish. Here are recipes for two classic sauces.

ROUILLE

This is the fiery mayonnaise of Provence, eaten atop croutons in bouillabaisse. But it has a hundred uses – with crudités for example, with globe artichokes and asparagus. It is one of the great sauces. Add 1 teaspoon of the above harissa to a mayonnaise made with 250 ml / ½ pt olive oil. Or follow the recipe below.

1 large dried red chilli
1 garlic clove, crushed
225 ml / 8 oz olive oil
2 egg yolks
pinch of sea salt

Blanch the chilli and the garlic and leave for an hour. Peel the garlic, slice the chilli and place in a liquidizer with a little of the olive oil – enough to turn the chilli and garlic into a paste.

In a separate bowl add the oil drop by drop to the egg yolks. As they emulsify, continue to pour until all the oil has been absorbed. Lastly, add the chilli paste and the salt. Beat thoroughly.

Some cooks suggest adding a drop of tomato purée for colour; it doesn't do the flavour too much harm either, but it is strictly not a classic rouille.

Short-cut rouille: Add to the aïoli mayonnaise 1 teaspoon Tabasco sauce and 1 teaspoon of hot red Hungarian paprika or Spanish *pimentón*.

HARISSA

This is the North-African sauce traditionally served with couscous.

115 g / 4 oz dried red chillies
6 garlic cloves, unpeeled
1 tsp ground coriander
1 tsp caraway seeds
1 tbsp dried mint
handful of fresh coriander leaves, chopped
2 tbsp olive oil
1 tsp sea salt

Soak the chillies in enough warm water to cover for an hour. Blanch the garlic and when cool peel its outer skin away.

Place the chillies with their soaking water into a blender with the garlic, coriander, caraway and mint. Liquidize to a purée, add the coriander leaves, oil and salt. Liquidize again.

The sauce keeps happily in the refrigerator for weeks if covered in oil.

RIGHT *Grilled peppers and anchovy (page 203) and rouille served with crudités*

Umbelliferae

THE PARSLEY FAMILY

CARROT *Daucus carota*
CELERY *Apium graveolens* var. *dulce*
ALEXANDERS *Smyrnium olusatrum*
CELERIAC *Apium graveolens* var. *rapaceum*
FENNEL *Foeniculum vulgare* var. *azoricum*
PARSNIP *Pastinaca sativa*
SKIRRET *Sium sisarum*
PARSLEY *Petroselinum crispum*
HAMBURG PARSLEY
Petroselinum crispum var. *tuberosum*
ROCK SAMPHIRE *Crithmum maritimum*

This family includes about 250 plant genera across the temperate regions of the world. Many are aromatic herbs (anise, dill, caraway, chervil, lovage, angelica and cumin) with feathery leaves, but all have the flowers arranged in a conspicuous flat cluster (which is sometimes slightly dome-shaped) at the top of their stalks – this inflorescence is the umbel that gives its name to the family.

*Many species of umbelliferae are poisonous; the family includes hemlock (*Conium maculatum*), water hemlock (*Cicuta maculata*) and fool's parsley (*Aethusa cynapium*). But other species are household* *names which we happily consume all the time. The aromatic leaves are consumed in parsley; the tap root is edible in carrot and parsnip (and, in its swollen form, in celeriac), while the succulent stems are the attraction in celery and fennel.*

CARROT

Daucus carota

One might think that carrots, like cabbage, radish and garlic, have a long and detailed history, but what little was written about them is both obscure and conflicting. Surely this phallic root with such a luscious colour would be admired, praised and celebrated in antiquity? But the truth is that the carrot as we know it is a much later addition to the table, while the wild carrot that the ancients knew was, and is, small, pale and tough. The yellow, lilac or purple carrot came from Afghanistan and is unlikely to have travelled to classical Greece until Alexander brought it back from Persia, and this root, I suspect, though purple, must also have been small and irregular in shape. The orange-red carrot, grown large, would have been nature's gift to the Greek phallic cult; even the small purple carrot might have had its devotees, yet neither is ever mentioned. In the whole of Athenaeus, though much is written about other vegetables, the carrot (and its cousin, the parsnip) is never alluded to.

The cultivated form of the purple carrot could have appeared in Asia Minor in the eighth century BC. It is thought to have been recognized as part of the gardens of Babylon, but it is placed among the scented herbs of King Merodach-Baladan (d.694 BC). Pliny mentions a Syrian plant called '*daucon*' in Greece and '*gallicam*' in Italy. It could be a carrot or a parsnip; he describes it as 'slender and bitter', which makes it quite unlike our present carrot with its high sugar content. Columella mentions carrots briefly, while Apicius suggests that they be eaten raw with salt, oil and vinegar and gives one modest recipe for cooking them in cumin sauce with a little oil.

By the second century AD, Galen comments that the wild carrot is less fit to be eaten than the domestic. So we at least now know that the carrot was being cultivated then. But was it our carrot? Surely, if so, its colour would have been remarked upon? No; Galen must be talking of the purple variety. Even if the Romans seemed apathetic, you would think that, as the Dark Ages closed over Europe and most of the peasantry were reduced to grubbing for roots, the sweetness of the carrot would make it both precious and popular. But it is ignored. Surely this is proof enough that today's carrot, which we know and love, was not yet formed and that what was left at the end of the Roman civilization was nothing remarkable, something not far removed from the wild carrot itself. Yet the carrot responds to cultivation and speedily improves. To breed our red carrot, you would need the yellow variety to be crossed with the purple and this did not seem to occur for centuries, though both were indigenous to Afghanistan.

Albertus Magnus (1200–80), the great German scholar, who was particularly precise and detailed, mentions the carrot's cousin the parsnip, but fails to record the carrot, which suggests that the vegetable may have been used in its wild state, but was still uncultivated in the thirteenth century. Both purple and yellow carrot seeds were brought by the Moors to Spain and because of the Spanish conquests in Flanders the vegetable was grown in the Netherlands by the late fifteenth century. From there it reached England and the feathery tops became a great success as decoration on hats and clothes in Elizabeth's court.

Perhaps we can narrow down the date of one step towards the true cultivation of today's carrot, for the first mention of it in a way we can recognize comes in the instructions that the Goodman of Paris compiled for his sixteen-year-old wife. He tells her that the carrots are purchased in the market in bundles and though red they always include one white one. Now the Goodman is particular about his kitchen garden and grows beets, leeks, cabbage, parsley, peas, spinach, lettuce, pumpkin, turnip, radish and parsnip among the vegetables. But the carrots have to be bought at market and they are not yet uniform in colour. The Goodman's book was written in 1392; Albert Magnus had written his work around 1260, by which time the Moors had already introduced into

Spain sugar cane, rice, cotton and the citrus fruits.

Recipes for carrot, and quite often for orange, are legion in Moorish cooking and they, in turn, were influenced by the Persians. Margaret Shaida (*The Legendary Cuisine of Persia*, 1992) claims it was the early dark purple carrot that the Persians cooked with and it was this carrot that Alexander the Great took back to Greece after he had conquered the Persian Empire. If so, as she admits, it never really made an impact on either Greece or Rome.

However, when the carrot reached England it was still thought of as either a pot-herb or a medicinal aid. Gerard mentions a yellow variety but tells us that the purple carrots were still the most popular, even though they lost their colour when cooked and turned brown. One suspects that this was the reason the Romans were less than enthusiastic, since they had an aesthetic eye for what looked good upon the table. By the early seventeenth century in Italy, Castelvetro was preparing salads from pink and yellow carrots, with pepper as the most important seasoning. But it was in Holland in the seventeenth and eighteenth centuries that carrots were bred in orange and red varieties. They were bred for sweetness as well as colour (though the sweetness must have taken time, for Ude in 1824 recommends blanching carrots 'to take off the tart taste'). Though white and purple varieties were still popular in France at the beginning of the nineteenth century, Dumas recommends the best carrot as being palish white or yellow; as his *Dictionary* was published in 1873, this shows a quite late disdain for the red carrot. Dumas gives a recipe for carrot soup, but one feels that the colour might have been too pale to have been appealing. Yet this was the century when carrots came into their own in France, associated with the spa, Vichy, where raw carrots were given daily as part of the cure, or Crécy, where the chalk soil is supposed to be good for carrot cultivation. In fact, carrots like a light loamy soil, which encourages a long straight root.

NUTRITION

Carrots are rich in carotene and vitamin A and contain significant amounts of B3, C and E. They also contain sodium, calcium, phosphorus, potassium, fibre and folic acid. All the nutrients are stored close to the surface, so it is best to wash or scrub carrots, but not to peel them.

Research ties regular consumption of beta-carotene vegetables and fruits with a defence against lung cancer. The myth about helping night sight – RAF pilots in the Second World War were encouraged to eat carrots – has elements of truth in it, as vitamin A is essential for the correct functioning of the eyes. Carrot juice is also used to prevent eye, throat, tonsil and sinus infections.

CHOOSING

The most delicious carrots are the young thinnings, about the size of the little finger. Most people would choose to eat them raw, but if cooked they need only to be thrown in a little boiling water for a minute.

Try and pick carrots loose or, better still, in bundles with their feathery tops looking fresh and green, but not packaged in polythene. Feel them for firmness. Old, tired carrots are best avoided altogether; their nutritional richness declines seriously, so they have little to offer. This is not to say that the mature, largish carrot, newly dug or bought fresh, is not a marvellous vegetable, scrubbed lightly, then used for a myriad of different dishes. Grated raw the flavour is fine, sprinkled with celery salt and with lemon juice added at the end. Or use in any of the following ways.

PREPARING AND COOKING

Wash and trim but do not peel the carrots. Their cooking times depend on their size and how they are cut. Whole carrots need about 20 minutes boiling or baking. You can bake them like potatoes in Gratin Dauphinois (see page 186), sliced thinly, or in stock with garlic. You can roast them, chopped in chunks, around a joint or game bird.

An excellent simple carrot soup can be made from stock and half a dozen carrots, boiled and then blended. The classic Crécy soup includes onion, butter, potato, egg yolks and cream but I have never found such additions necessary or even worthwhile. You might, though, like to experiment with adding orange zest to the carrots when they begin cooking and some orange juice at the end.

Another simple recipe is all the variations on carrot salad, where the vegetables are grated and mixed with oil and vinegar and perhaps other ingredients and flavourings. Again, you could add orange juice – or even rose water in honour of the carrots' early Persian past. Such salads go well with roasted pine nuts, or roasted mustard seeds. The carrot, after all, is also much used in Indian cooking and there are many raw salads with mixtures of spices which are both exciting and satisfying.

One of the classic ways of cooking carrots is to add butter to a little of the poaching water and to finish with some sugar so that the vegetables absorb the liquid and end up glazed. It is pretty to look at and highly satisfying. I have rarely known carrots cooked in this way to be left. This method goes by various names. In the late 1800s it is referred to as THE FLEMISH WAY. The carrots are simmered in about 30 g / 1 oz of butter, a wine glassful of water, salt and pepper and then finished with a pinch of sugar, a dash of chopped parsley, two egg yolks and a little cream. Elizabeth David has a less rich version called CAROTTES VICHY, where the carrots are cooked with 30 g / 1 oz of butter, salt and two lumps of sugar, then finished with more butter and parsley.

My own version dispenses with the parsley altogether and uses the same amount (30 g / 1 oz) of butter (to, say, 450 g / 1 lb of carrots sliced diagonally) and 150 ml / ¼ pt of water. When the water and butter are absorbed, add a little more butter and a heaped teaspoon of muscovado sugar. Shake the pan for 2 minutes over the heat so that the carrots become glazed and do not stick. Instead of adding sugar at the last moment, try adding marsala, madeira or a dessert sherry. Carrots also go well with a little finely sliced root ginger, added before the sugar or wine.

Do not always grate carrots when making salads. Slice them paper-thin with a mandoline or in a food processor, then dress in the Indian manner: fry a selection of seeds – mustard, cardamom, cumin and coriander – in sesame or mustard oil, then pour them over the raw carrots with the addition of lime and lemon juice, salt and pepper.

The idea of using carrots as a savoury is as recent as the last century, for carrots were used since ancient Persia in sweet dishes. Now, Margaret Shaida tells us, carrot 'is conserved and preserved in jams and syrups, partnered with orange peel and almonds.'

In the Second World War carrots were used extensively in jams, puddings and cakes. It seems to be only in the latter that we still use them. This recipe for carrot cake is far and away the best I've ever tasted. It was published in Michael Bateman's column in the *Independent on Sunday*; he thinks he got it from the book *Laurel's Kitchen*. I have added the alcohol and the orange zest and juice, though lime will do. It is extravagant with honey, but that is the secret of its extreme deliciousness.

CARROT CAKE

300 ml / 10 fl oz honey
juice and zest of 1 orange
wine glass of dark rum
110 ml / 4 fl oz water
115 g / 4 oz carrots, finely grated
115 g / 4 oz butter
115 g / 4 oz stoned raisins
85 g / 3 oz chopped dates
1 tsp ground cinnamon
½ tsp grated nutmeg
½ tsp ground cloves
225 g / 8 oz wholewheat pastry flour
pinch of sea salt
2 tsp bicarbonate of soda
115 g / 4 oz shelled walnuts
for the topping:
150 g / 5 oz cream cheese
150 g / 5 oz crème fraîche
2 tbsp softened butter
3 tbsp icing sugar
1 tsp vanilla extract

In a saucepan, heat the honey, orange zest and juice, dark rum and 110 ml / 4 fl oz water with the grated carrots, butter, raisins, dates and the ground cinnamon, nutmeg and cloves. Boil for 5 minutes, then leave to cool until lukewarm.

In a bowl, mix the flour with the salt, then add the bicarbonate of soda, and finally the walnuts. Make

ABOVE Carrot cake

a well in the middle, and pour in the carrot mixture, blending well.

Preheat the oven to 180°C/350°F/gas4 and bake in a greased cake tin (25 cm / 10 in across) for about 1 hour. The cake is done when you can insert a skewer and it comes out clean. Leave it in its tin for 10 minutes before turning it out on to a wire cake rack.

For the topping: cream the cheese with the crème fraîche, mix in the softened butter and beat in the icing sugar and the vanilla. Spread the topping over the cake when it has cooled.

CARROT PUFFS

This is an eighteenth-century recipe quoted in Dorothy Hartley which I adjusted and placed in my first cookery book: 'Scrape and boil them and mash them very fine, add to every pint of the pulp about ½ pint of bread crumb, some eggs, four whites to the pint, a nutmeg grated, some orange flower water, sugar to taste, a little sack, and mix it with thick cream. They must be fried in rendered suet very hot.'

That is an eighteenth-century English recipe; the following one is a modern version. These puffs are far more delicious than they might sound.

Boil 450 g / 1 lb sliced carrots; drain, then mash them. Add the juice from 1 orange, the yolks of 2 eggs and enough breadcrumbs to bind the mixture. Then fold in the whipped whites of the eggs.

Have a little olive oil in a pan, and when it is hot enough, mould the mixture with your fingers into round puffs the size of a ping-pong ball. Drop these into the oil and roll them around in the pan until they are brown and crisp on the outside. (They should be crunchy on the outside and light in the centre.)

Pile them on a dish and serve them at once.

CARROT APPETIZERS

This carrot pickle is derived from one of Julie Sahni's recipes in her *Classic Indian Vegetarian Cooking*.

3 tbsp mustard oil
1 tbsp mustard seeds
1 dried red chilli, broken up
½ tsp turmeric
½ tsp asafoetida
450 g / 1 lb carrots, cut into sticks
½ tsp salt
½ tsp sugar
2 tbsp lemon juice

Heat the oil and cook the seeds, chilli, turmeric and asafoetida. Add the carrot sticks, the salt and sugar. Stir-fry for 2 minutes, then add the lemon juice and cook for another minute.

CARROT AND RED PEPPER TERRINE

450 g / 1 lb carrots
2 tbsp olive oil
3 large red peppers, cored, seeded and sliced
1 dried red chilli, crushed
115 g / 4 oz ricotta cheese
115 g / 4 oz strong Cheddar cheese, grated
3 eggs, beaten
1 tbsp Dijon mustard
½ tsp asafoetida
1 tsp sea salt

Trim and dice the carrots, so they are about the size of a thumb nail, throw them into 1 tbsp olive oil and sweat for 5–10 minutes, then reserve.

Heat the remaining olive oil and cook the sliced peppers and crushed chilli in it in a closed pan over a low flame for 20 minutes.

Leave to cool, then add the ricotta cheese, the eggs, mustard, asafoetida and salt.

Line a terrine dish with well-buttered foil with enough foil overlapping to use as handles. Pour in the terrine mixture, cover with buttered paper and place in a bain-marie in a preheated oven at 220°C/425°F/gas7 for 50–60 minutes, or until it has risen.

Take out and leave for 1 hour before unmoulding, and refrigerate before serving. Serve on a bed of rocket or delicate salad leaves, accompanied by pesto, the celery sauce on page 220, a bitter orange sauce (made in winter, when Seville oranges are in season), or a laver sauce. This last is made from laver bread – seaweed – available ready cooked (and of excellent quality) in tins. Simply empty it into a pan, mix with a little butter or orange juice, and reheat.

CELERY

Apium graveolens var. *dulce*

I have eaten wild celery (*Apium graveolens*) in both Greece and Sicily and you will find smallage, which is a cultivated wild variety, piled high in the markets of Italy and France. There they consider smallage by far the best type of celery to cook with, flavouring soups and stews with that very distinct aromatic savouriness. 'It is an agreeable pursuit to roam the pleasant, wild countryside, picking plants here and there... should you find lovage or mountain celery – to be picked green, fresh and fragrant.' The fourteenth-century herbal *The Four Seasons of the House of Cerruti* goes on to warn that the smell makes the head feel heavy and that it heats the blood.

Wild celery is certainly much more pungent in smell and taste, but also naturally a great deal more fibrous. The stalks are thin and very green, but it is a pleasure on a walk to pluck the stems and suck the juice from them. I have never noticed a heaviness in the head as a result, but the plant is a diuretic, so you may have to halt here and there.

The wild plant, like so many of our food plants, grows in marshy salt-impregnated ground near the sea shore. It is a tough little plant extending from Sweden to Egypt, from Britain to Asia and has even been found in New Zealand and Tierra del Fuego. The ancients used it medicinally and for flavouring, yet also considered it a plant of funereal disposition. One type of celery is mentioned as standing as the doorkeeper to heaven; it was woven into garlands to adorn mummies in Egypt. Sturtevant reviews the evidence from the Ancient World and considers there is not one shred of evidence for thinking that celery was used as food then or later. Yet Athenaeus mentions celery – obviously the wild kind, for he says it is similar to marshwort and likes to grow in water. We have a complaint by an Attic playwright, Epicharmus, that devouring grubs wind their way 'in and out among the leaves of basil, lettuce and fragrant celery.' Which certainly implies the plant was cultivated in a Greek garden. If celery was this wild variety called smallage and if it was extensively used in the kitchen, its seeds were certainly prized as much as its stalks. Pliny writes of the pleasure given by 'celery stalks swimming in broth'. These must have been young shoots which would have been tender.

Apicius cooks it with green beans, leeks, rue, oil and liquamen (see page 143), he reduces it to a purée with lovage, oregano, onion and wine, he uses it to stuff a suckling pig and in a vegetable stew with bulbs. He also suggests it makes a good laxative.

Luigi Alamanni, who wrote *La Coltivazione* (1546), six books in praise of the rustic life, mentions celery in passing but praises the quality of Alexanders for the sweetness of its roots (see overleaf). However, Castelvetro sixty years later gives detailed directions on how to grow the plants and how to blanch them in early autumn. This is obviously still smallage, for though he says the plants need to be planted out seven inches apart 'for they grow quite large heads', today's celery would need twice that space. Castelvetro is an exception in that he recommends eating the celery, for though many horticulturalists speak of its cultivation and of the wild plant being transferred to gardens, it is, they stress, not for food but for medicinal use, while the seeds only might be added for flavouring. It was thought to purify the blood. Cultivated smallage in France is called *céleri à couper* (celery for cutting and not necessarily for digging up, also called *petit céleri*) and is used in soups, stews and broths.

The celery we now know did not appear until the eighteenth century. Obviously breeding it was a struggle, for John Ray mentions that in English gardens the cultivated form often degenerates back into smallage. By 1778 two sorts of celery are noted in England, one with hollow stalks and the other with the stalks solid. In Sweden, the richer classes preserved their celery in the cellars for their winter use.

Whether smallage or cultivated celery, the distinct flavour had its passionate adherents. Used for centuries in soups and broths, by the beginning of the nineteenth century celery as we know it now was a firm favourite among vegetables in the kitchen.

Ude (cook to Louis XVI and then the English aristocracy) in 1824 gives three recipes – a purée, stewed celery with white sauce and celery *à l'espagnole*, made with that Spanish sauce which had so recently swept France with its addictive popularity. The sauce is made with a great quantity of Serrano ham. One recipe for Spanish sauce required 'twelve ducks, a ham, 2 bottles of old Madeira and 6 pounds of fine truffles'. It was Louis XV's cook, Menon, who made

BELOW *Havesting celery*

Spanish sauce a success in France, as he always put a double amount of ham into it, which predominated over every other ingredient.

Mrs Beeton, a little later in the century, gives two recipes for celery sauce, the first flavoured with spices (2 blades of mace might be overpowering) and herbs and finished with cream and lemon, the second without the herbs and cream. Mrs Beeton also gives celery salad (which mixes the raw vegetable with either Stilton or Cheddar, both being 'masked' with mayonnaise), a celery soup, and three recipes for stewed celery. She ends with one for Celery Vinegar where ¼ oz of celery seed is pounded in a mortar, 1 pint of vinegar is boiled, then when cold poured on to the seed, it is left for 2 weeks to infuse, then strained. 'This is frequently used in salads,' she says.

From then on celery was widely cultivated and was used extensively in the kitchen. In the 1930s no banquet would have been complete without braised celery as one of the vegetables. Celery soup was also a favourite at dinner parties and from the same period Arabella Boxer quotes a celery sauce to be served with game. Sadly, we seem to have lost the art of making soup from celery (see below) though, unfortunately, tinned celery soup seems to be always with us, like the ghost of cuisine past.

ALEXANDERS

Smyrnium olusatrum

So called because it is supposed to have originated in Macedonia, the kingdom of Alexander the Great, but if so it also spread itself over Eastern Europe and Asia Minor. It is eaten with enjoyment in the Ancient World, where the raw root was used as a salad, much as we now use fennel. In the USA in the nineteenth century Alexanders was blanched by earthing up the stems and cooked and eaten like cardoon.

In between those years it also enjoyed much popularity. It was mentioned by Theophrastus in 322 BC and then by Pliny and Columella. The Romans brought it to England as one of their pot herbs and it remained part of the kitchen garden until the eighteenth century. It appears in Charlemagne's list of vegetables to be grown on the estates of his ninth-century empire. Gerard in 1597 says it 'groweth in most places of England'. It still does, growing in colonies near the sea. A biennial, it is easy to recognize by its greenish-yellow umbelliferous flower head.

The whole plant is edible, though both leaves and stalks become impossibly tough when mature by midsummer. When young, the stalks are delicious steamed or boiled. In *Wild Food*, Roger Phillips counsels peeling them like rhubarb, and even on youngish plants the knuckly leaf-joints may have to be cut off. The Greek name *smyrnium* derives from myrrh and refers to its scented, myrrh-like taste. The flavour is stronger than that of celery, highly aromatic and with a tinge of aniseed. Other names are 'black lovage' (its black seeds were sold by apothecaries under the name 'Macedonian parsley') and 'horse parsley'.

This is an important food plant which we now seem almost to have lost in cultivation. A few seedsmen list Alexanders under herbs, but enterprising gardeners could gather a few of the plentiful wild seeds to sow in their own gardens.

NUTRITION

With a vegetable as delicious as celery it is somewhat of a surprise to discover that there is little food value in the plant. But celery does raise the alkali levels in the body, so it is excellent for people who suffer from ailments like gout and rheumatism. It is also helpful for high blood pressure. It is a perfect food for slimmers, as it has no calories to speak of, and a few sticks chewed before meals can adequately take the edge off the appetite.

VARIETIES

The main difference is between the green or yellow 'self-blanching' types that are available in supermarkets all year – home-grown in summer and imported at other times, and the frost-hardy white celery that is available only in winter. Gardeners can also buy smallage seeds as 'cutting celery' from some of the seedsmen who cater for minority tastes.

RIGHT *Celery and celeriac*

CHOOSING, PREPARING AND COOKING

I cannot pretend to any liking for the celery that is packaged in supermarkets; the taste is so inferior, it seems to me no pleasure at all to eat. The green celery which comes from California and Spain, which is also enjoyed in Italy, is all water and no taste. I want what they call around me in Suffolk 'dirty celery', that is, celery which has been earthed up to be properly blanched.

This celery has the right nutty bite and is also worth cooking with as well as eating raw. The season is from November to February, too short for this wonderful vegetable. But if you see it on sale in country markets, buy several heads, for it will keep well, as long as it is undamaged, in a dry, cool dark place.

To prepare, break off the outer stems which are too fibrous or damaged to eat raw and scrub the dirt off with a brush under running water. Keep off-cuts and leaves for flavouring stocks and so on. Have a large jug ready filled with cold water and as you clean the stems put them in that. Shave the rough bits of celery off the core or heart, for this has the best flavour and leave it attached to the inner stems. Eat raw dipped into sesame salt or use for cooking.

Celery also enjoyed in the 1950s and '60s a great vogue as an appetizer to be served with the drinks at parties. They were called 'celery boats' and only the stems with deep troughs were chosen. These were filled with a mixture of blue cheese and butter, or cream cheese and walnuts and either left whole, in which case the guest waved a baton of stuffed celery around, or else sliced into bite-sized pieces.

CELERY SOUP

2 heads of celery, trimmed
30 g / 1 oz butter
½ tsp white pepper
½ tsp celery salt
1.75 litres / 3 pt vegetable stock
300 ml / ½ pt single cream or smetana
2 – 3 tbsp finely chopped leaf coriander

Wash the celery thoroughly, cutting off and throwing away all damaged parts, then slice the rest across. Melt the butter in a large pan and sweat the celery with the pepper and celery salt.

Add the stock and simmer for 20 minutes, leave to cool and then blend. There should be no fibres if only the inner stems have been used and it has been cooked enough. But if there are, put the soup through a coarse sieve.

Reheat gently, adding the single cream or smetana, then sprinkle the chopped leaf coriander on the surface of the soup before serving.

For *Celery Sauce* use the same method but add no stock. Let the chopped celery cook in the butter in a closed pan over a low heat for half an hour, then leave to cool and liquidize, adding a small wine glass of dry sherry and a tablespoon of double cream.

BRAISED CELERY WITH MUSHROOMS

2 heads of celery
30 g / 1 oz butter
1 tbsp olive oil
15 g / ¼ lb mushrooms, sliced
½ tsp asafoetida
150 ml / 5 fl oz white wine
1 tsp plain flour
sea salt and freshly ground black pepper

Discard the damaged outer stalks of celery, then cut the leafy top half of the stalks away, leaving the bottom half and the heart. Slice each head in two and wash thoroughly, turning the heart upside down and rinsing under running water.

Melt the butter in a shallow oven dish, add the olive oil, mushrooms and asafoetida, stir in the white wine and let it all bubble and cook for a moment. Season, then lay the 4 celery halves, flat side down, into the sauce.

Place in an oven preheated to 190°C/375°F/gas5 for 30 minutes. Take out of the oven and thicken the sauce with the flour.

CELERIAC

Apium graveolens var. *rapaceum*

'There is another kind of celery called Capitatum, which is grown in the gardens of St Agatha, Theano and other places in Apulia, granted from nature and unseen and unnamed by the Ancients. Its bulb is spherical, nearly the size of a man's head. It is very sweet and odorous...'. So writes Baptista Porta, a Neapolitan, in his *Villae* published in Frankfurt in 1592. Porta goes on to say that if the soil is not rich then the plant will degenerate and differ little from the usual smallage.

Where did such a close cousin of celery derive from, with its massive rounded roots (*rapaceum* means 'turnip-like')? Porta was writing from the kingdom of Naples, which covered southern Italy. Apulia (now Puglia) was dotted with ports which had a thriving trade with Asia Minor and the Eastern Mediterranean. A little earlier, around 1573–5, a traveller in the East, Rauwolf had spoken of '*eppich*', a root that was a delicacy eaten with salt and pepper which he had tasted at Tripoli and Aleppo. Smallage had far too small and undistinguished a root to be treated as a delicacy; besides, it is too bitter to be eaten raw, so Rauwolf's observation is thought to be the first mention of celeriac or celery root. But it is not until 1729 that the plant is described in England by Stephen Switzer, a writer and seedsman in Westminster Hall; he admitted that he had never seen it, but he had been furnished with a supply of seeds for the plant from Alexandria.

Alas, though Switzer wrote a pamphlet 'Growing foreign vegetables', given away with the seeds, celeriac did not catch on in England, though it made some headway in both France and Germany. Certainly in France it gave its name to one particular dish, *celeriac rémoulade*, which I always think is overrated. There are better things to do with raw grated vegetables than cover them in rich mayonnaise. Another classic way of treating celeriac is to purée it and mix this with the same amount of puréed potato. Many writers are ecstatic about this dish, but why, I ask myself, dissipate the fine flavour of celeriac with the purée of potato so that you have neither? The flavour of celeriac is too fine and subtle for such an insensitive treatment.

It must have been Elizabeth David in *French Provincial Cooking* who first alerted the British to the existence of this vegetable. She it was, too, who first suggested the mixture of potato and celeriac but on the same page she gives a simple and delicious recipe for celeriac stewed in butter (see below).

NUTRITION

Astonishingly, unlike celery, celeriac is a storehouse of energy, for the growing plant is high in carbohydrate and minerals, it also has vitamin C and some B vitamins and is very rich in iron.

PREPARING AND COOKING

Buy the smaller bulbs of celeriac, about the size of a tennis ball or a little larger and buy them when the green leaf is vivid in colour and has not faded. Larger celeriac bulbs tend to go woolly in the centre and most of it then should be thrown away. Choose smooth-skinned bulbs for less waste. Peel the knobbly outer skin away and use the celeriac at once. If left, the flesh will oxidize and brown and look unsightly, if you do have to leave them for a few minutes, pop them in acidulated water.

One of the simplest methods of cooking celeriac roots is to slip them cut into chunks the size of a roasting potato into the roasting pan with the meat juices and let them roast for half an hour. If made into a purée like swede they tend to be a little watery, which is why, I surmise, the idea of mixing them with mashed potato came about. But they can be substituted as the vegetable in all the potato gratin dishes (see page 186).

Cream has always been an acceptable foil, though too rich, I suspect, for our tastes today. I tend to use them in ways where they do not combine with cream or mayonnaise. Once cut or sliced, celeriac cooks very quickly, far more quickly than celery. Mrs David's timing for the following recipe – 10 minutes – is, in my opinion, a mite too long, but such things depend so much on the heat of the ring and are impossible to prescribe precisely.

CELERIAC *Apium graveolens* var. *rapaceum*

CELERIAC STEWED IN BUTTER

2 medium celeriac roots
30 g / 1 oz butter
2 tbsp olive oil
1 tsp Dijon mustard
a few drops of wine vinegar
sea salt and freshly ground black pepper
a little finely chopped parsley

Peel the celeriac and slice thinly on a mandoline or in a food processor.

Heat the butter and olive oil in a wok or frying pan and throw in the celeriac slices. Cook them, moving them around, over quite a fierce heat for about 5 minutes, then add the seasoning, mustard and wine vinegar. Stir-fry for another 2 minutes, when they should be just browned at the edges.

Serve sprinkled with the parsley.

CELERIAC AND HALUMI SALAD

1 packet of halumi cheese
2 medium celeriac roots
2 tbsp olive oil
2–3 garlic cloves
1 dried red chilli, crushed
2 or 3 stems of celery heart
lemon or lime juice
sea salt and freshly ground black pepper
chopped leaf coriander or finely sliced spring onion

Drain the halumi cheese and cut the slab into 1 cm / ½ in cubes, throw into a frying pan that is *without fat* and dry-fry until each cube begins to melt a little, turning them over so that each side browns. This will take a little longer than you think – about 5 minutes. When done, throw into a bowl and reserve.

Peel and trim the celeriac, then dice into about the same size cubes. Heat the olive oil in a frying pan, add the garlic and chilli, then throw in the celeriac, season with the salt and pepper and cook until the celeriac is just shaded with gold.

Throw into the bowl with the halumi. Slice and add the celery, squeeze the lemon or lime juice over the lot, mix thoroughly and sprinkle with either coriander or spring onion. Makes an excellent starter served on a bed of salad leaves.

WARM CELERIAC SALAD

2 medium celeriac roots
2 garlic cloves, crushed
3 tbsp walnut oil
1 tsp wine vinegar
1 tsp Dijon mustard
½ tsp sugar
sea salt and freshly ground black pepper
a little chopped parsley and spring onion

Peel and trim the celeriacs, then slice them across in 5 mm / ¼ in chunks. Slice these down into large chip size. Steam these for 5 minutes.

In the meantime, make the vinaigrette by mixing all the rest of the ingredients. Toss the warm celeriac pieces into the vinaigrette and serve immediately, either as a first course on a bed of salad leaves or as a salad after the main course.

STUFFED CELERIAC

2 medium celeriac roots
2 or 3 tbsp olive oil
55 g / 2 oz root ginger, peeled and grated
1 red or green chilli, finely sliced
115 g / 4 oz mushrooms, sliced
115 g / 4 oz oyster mushrooms, sliced
1 red or orange pepper, cored, seeded and sliced very thinly
wine glass of madeira, marsala or dessert sherry
sea salt and freshly ground black pepper
a little coriander or spring onion, finely sliced

Peel and trim the celeriac roots, cut them in half and square off each bottom so that they sit firmly, cut side up. Hollow out a circular portion in each half of celeriac about 1 cm / ½ in deep; and 5 cm / 2 in in

RIGHT *Celeriac and halumi salad*

diameter (though this depends on the size of the original root). Chop the pieces taken out and reserve.

Pour the oil into a pan and brown the 4 scooped-out halves, then remove and place them in an oven dish and bake at 190°C/375°F/gas5 for 30 minutes.

In the meantime, throw the ginger root and chilli into the oil, followed by the chopped celeriac and the mushrooms. Toss until they are almost cooked, about 4 minutes. Then add the sliced pepper and cook for another 2 minutes. Pour in the madeira and season. Mix thoroughly, then take off the heat.

Spoon this mixture into the 4 celeriac containers, garnish with the coriander or spring onion and set on a bed of leaves. Serve warm, though it is unexpectedly good when cold.

FENNEL

Foeniculum vulgare var. *azoricum*

Florence fennel, sweet fennel or *finocchio dulce* is the white bulb that we eat raw or cooked; in Italy in the past it was also served as a dessert, with the fruit and cheese. There is no difference in flavour between the bulb, the leaves and the seeds – all taste very pleasantly of aniseed in varying degrees of strength. There is also little difference between the plant we use as the herb and the one whose bulb we eat. So it is often difficult to distinguish in the past which fennel is being eaten.

Like many gardeners before me, I once attempted to grow the bulb fennel and found it turned into the herb fennel. No one had told me that the bulb has to be harvested very early in its growth; the stems must not grow more than a foot high, for the bulb is in fact the expanding leaf stalks, and once they grow above ground there is no bulb to speak of. The plant is also difficult to grow in Britain's climate; it needs warmth and rain to begin with, and while having plenty of the latter, we seldom have the former at the same time.

Fennel as a herb was certainly one of the first to be cultivated in the early Assyrian and Babylonian gardens (others were cumin, sesame, mint, basil, coriander, anise, thyme, asafoetida, bay, rocket, saffron and sage). All of these and more were later grown by the Greeks and Romans. As a seasoning, fennel was one of the most popular herbs in the Ancient World. The new stalks of fennel were considered a great delicacy and were earthed up or blanched to make them tender. Columella gives a recipe for preserving stalks and shoots – among them cabbage sprouts and fennel – in brine and vinegar. He also uses fennel seed with toasted sesame, anise and cumin mixed with puréed dried fig and wrapped in fig leaves and then stored in jars to preserve it.

Pliny regards fennel as being very useful 'for seasoning a great many dishes'. He also makes the curious observation that snakes are very fond of the plant. Archestratus (the fourth-century BC Greek gourmet) tells us that fennel was placed in the brine with olives when curing. Indeed, it appears in nearly every herb mixture.

Fennel seeds were immensely popular from the earliest times, used both medicinally and as flavouring. They were constantly mentioned in Anglo-Saxon medical recipes and were a favourite plant of Charlemagne. Fennel shoots, fennel water and fennel seed are all mentioned in an ancient record of Spanish agriculture of AD 961. John Evelyn in *Acetaria* talks of the 'sweetest fennel of Bolognia' which is aromatic, hot and dry. He considers it 'expels wind, sharpens the sight and recreates the brain; especially the tender Umbella and Seed-pods.' Evelyn recommends eating the stalks peeled and dressed like celery and tells us that the Italians eat the blanched stalk, which they call *cartucci*, all winter long. But did they know and consume the bulb? Did the Romans themselves grow it? Sturtevant considers '*finocchio*' in the form of sweet fennel to be a late development and points to Stephen Switzer's mention of it in 1729 as having just been introduced to England, but it was still rare in 1765. It appears a little later, in a catalogue of 1778, as Azorian Dwarf or finocchio and again in 1783 as Sweet Azorian fennel, which surely implies that it was cultivated in the Azores and its seeds exported. However, it remained a rare vegetable, though the herb fennel, its seeds and leaves were used extensively as a sauce for fish.

Finocchio really only became a regular sight in our greengrocers within the last thirty years, though it has been used in Italy ever since it first appeared. Vincenzo Corrado gives eleven recipes for the bulb in his *Il Cuoco Galante*, published in Naples in 1778. Cooked in oil with anchovies, cooked in milk with cinnamon and grated nutmeg, cooked in capon stock, served with a prawn coulis, there seems no end to Corrado's invention. He must have loved the vegetable, and no wonder: it is amazingly refreshing, both raw and cooked.

The seeds are now used commercially in a multiplicity of ways, to flavour sweets, gum and liqueurs as well as being used in savoury and sweet dishes in the domestic kitchen. In India the seeds feature large in the flavouring of many dishes. Roasted seeds are often offered after the meal as part of a *digestif*.

NUTRITION

Fennel contains oil and protein, is high in vitamin A and E, calcium and potassium. Fennel and ginger make a good digestive tea. Fennel is low in calories and high in water content.

PREPARING AND COOKING

Buy the smaller bulbs as there is less wastage. If there are only large bulbs available, you can always boil the discarded tough outer skins for stock and soups, while the feathery leaves can be chopped and added as a fresh herb to sauces or stuffings.

Once the bulb has been trimmed of these outer tough leaves, it can be sliced whichever way you like – across in rings or downwards. I prefer the latter and simply serve the sliced bulb with oil and lemon as a separate salad. In a mixed salad it is a treat to discover a slice of the anise-flavoured root. It also looks good as part of a mixed selection of crudités. The sliced bulb is so attractive that to serve it raw any other way, for example grated, seems to me one of the great culinary crimes.

If you wish to use the leaves in fish cooking, they should be soaked in water first and then wrapped around the fish before the fish is baked or barbecued. It is a good thing to fill the gutted fish with the chopped leaves and stalks to further flavour it.

The cooked bulb goes well with Parmesan, tomato, olives and garlic, but it also makes a marvellously refreshing soup.

LEFT *Fennel niçoise (overleaf)*

FENNEL SOUP

4 heads of fennel, trimmed and chopped
30 g / 1 oz butter
1 large onion, sliced
1 tbsp fennel seed
2 litres / 3½ pt vegetable stock
300 ml / ½ pt single cream or smetana
sea salt and freshly ground black pepper

Reserve the green fronds of the fennel for garnish. Heat the butter and throw in the onion, fennel seed and sliced fennel. Cook for a moment or two, then add the vegetable stock and simmer for 20 minutes. Season to taste. Leave to cool, then blend.

Reheat gently, adding the cream or smetana. Before serving, garnish with the chopped reserved leaves.

LIGHTLY BATTERED FENNEL

This makes a tasty and unusual starter or snack.

2 heads of fennel, trimmed
oil for frying, preferably olive oil
for the batter:
115 g / 4 oz plain flour
1 egg
pinch of sea salt
150 ml / 5 fl oz water or milk, or half and half

First make the batter: stir the egg into the flour and salt, then add the water or milk. (If you use milk, the batter will be thicker and heavier; with just water, the batter is more like a Japanese tempura.) This batter can be used at once.

Slice the fennel down into 1 cm / ½ in pieces, dip them in the batter and fry in hot oil, turning once, until golden brown. Place on paper towels for a moment, then serve at once.

FENNEL IN A WHITE WINE SAUCE

2 or 3 heads of fennel, trimmed and quartered
a wine glass of dry white wine
15 g / ½ oz butter
1 tsp flour
sea salt and freshly ground black pepper

Reserve the green leaves and poach the quartered fennel in a little water, no more than 2.5 cm / 1 in deep, for about 5 minutes. Remove the fennel and reduce the water by half. Add the white wine and bring to the boil, season, then mix the butter and flour into a beurre manié and stir in to thicken the sauce. Pour the sauce over the fennel quarters, sprinkle the chopped green herb over and serve.

VARIATION

Place the fennel in its thickened sauce in an oven dish and sprinkle 30 g / 1 oz of grated Parmesan cheese over the top. Leave in the oven until the cheese has melted and the sauce is bubbling.

FENNEL NIÇOISE

2–3 tbsp olive oil
1 small can of anchovies
5 garlic cloves, sliced
3 or 4 heads of fennel, trimmed and quartered
5 large tomatoes, skinned and chopped
12–16 stoned black olives
freshly ground black pepper

Heat the olive oil in a pan and throw in the anchovies and garlic, then the quartered fennel. Turn the vegetables and cook for a moment, before adding the chopped tomatoes and the olives. Place a lid over the pan and leave to simmer for 15 minutes, when the fennel should be cooked. Season with black pepper.

PARSNIP

Pastinaca sativa

The wild parsnip came to Europe from the Caucasus, but it grew also in North America and Darwin found it in South America, growing around Buenos Aires. The root of the wild plant is aromatic, white and sweet, which is why it was so popular from the earliest times – although in the early records the parsnip is confused with the carrot, and one is never too sure which root is meant. Like all plants in the wild there is a great variety of types and sizes. At some point a fleshier-rooted form must have appeared.

The Emperor Tiberius had such a fondness for parsnips that he had them brought every year from Gelduba upon the Rhine, where they are said to have grown to perfection. Rome brought the cultivated parsnip to Britain and Gaul, along with carrots, turnips, radishes and skirrets. Pliny observed that both radishes and skirrets grew better in Britain and so, one imagines, did all these root vegetables. Certainly the parsnip was a feature in the crops grown by all the Anglo-Saxon farmers that came after the end of the Roman Empire.

The parsnip had many uses, not only as another root to be stewed in the one-pot meal but it was valued for its sugar content, it was crushed and the liquid drawn off was then boiled and used as honey. The whole root, like the wild parsnip, could also be used for fermented drinks and a wine or beer made from it. It was also used as a jam or marmalade and in sweet puddings and tarts.

Before the potato was introduced to the Old World (and certainly long before it found acceptance in culinary circles), parsnip was the main starchy vegetable to be served with the meal, roasted in the pan beside the meat or beneath it. John Evelyn's buttered parsnips were dusted with ginger for added sweetness. A popular Elizabethan dish was fritters of skirret and parsnip in butter sprinkled with sugar. But Evelyn, ever resourceful, also suggests boiling the parsnip and serving it cold as a winter salad with oil and vinegar. He adds that it is more nourishing than the turnip. Hannah Glasse gives a recipe for salt cod with hard-boiled eggs melted in butter and 'parsnips boiled beat and fine with butter and cream'.

Why are the French so dedicated in their dislike of the parsnip – *panais*? Did they not use it in the past? The Goodman of Paris sowed parsnip seed in his garden, yet the parsnip in France is used in *pot-au-feu* and little else. You will certainly never be served a dish of roast parsnips, for they always prefer turnips.

SKIRRET

Sium sisarum

Skirret produces clusters of roots similar in flavour to parsnips. An ancient plant, native to China, it was known in Europe in Roman times, and continued to be grown and eaten until superseded in recent centuries by other improved root crops such as carrots and parsnips. The sixteenth-century name, 'sister', was applied to the carrot as well as to skirrets; in France it was sometimes known as '*carotte blanche*'. Calling it 'sisarum', Evelyn describes it in his *Acetaria* as seldom eaten raw, but being boiled, stewed, roasted under the embers, baked in pies, whole, sliced or in pulp, is very acceptable to all palates'. He also commends it for not provoking wind. The fleshy tubers no bigger than a little finger are sweet and floury but perhaps something of a chore to clean.

NUTRITION

Parsnips are quite high in vitamin A, some of the B vitamins and vitamin C. They also have calcium, phosphorus and potassium, as well as plenty of other minerals. Of course, if they are boiled to a pulp, a lot of this nutrition is lost, but with circumspect cooking methods at least half is retained.

CHOOSING AND COOKING

Once parsnips have had a frost the starch in the roots turns to sugar, hence parsnips should always be eaten in midwinter. Gardeners can leave them in the ground and dig them when needed. Try to choose small parsnips; the larger ones tend to have a woody centre which has to be cut out. This feature is prominent in the wild parsnip and as they got larger the edible part got less. Small parsnips also need little to no peeling, but just trimming.

For roasting they need to be parboiled for 2 minutes – no more. But I prefer not to cook parsnips whole; cut them up into chunks about 8 cm / 3 in long, either halving or quartering the thick end, depending on how big the parsnip is.

After the parboiling, drain them well, then throw the parsnip chunks into the roasting pan around the bird or joint. They will be done in 30 minutes, but I confess to liking my parsnips and roast spuds well done – that is, turned a deep golden brown and highly crispy – so I have often left parsnips in the pan for twice that amount of time. They come out all crisp exterior with a little sweet gooey softness in the centre. Very satisfactory.

I must say that I am no devotee of the plain boiled parsnip, even with lashings of butter or cream. But glazed parsnips are quite another dish and these have been neglected in the repertoire of British cuisine, ever since medieval and Tudor cookery where sweetened parsnips featured as a favourite aspect of many meat dishes. The mainstream of parsnip cooking appears to be rural English (the Scots and Irish both prefer swedes and potatoes), where there are dishes many and various. Parsnips here are often, one suspects, a main dish baked with a little fat bacon, or boiled with some pickled pork. There are numerous recipes for parsnip cakes, fritters, soup and wine. In her *English Cookery Book* published in 1943, farmer's wife Lucie G. Nicoll tells us: 'The recipes are a selection of the best from many thousands I have encountered in twenty-five years of managing a farm household.' She goes on to talk of how many generations back some of these recipes go, for all were handed down. Overleaf I list a few – excellent, basic and tasty.

I have not included Mrs Nicoll's recipe for parsnip soup, as in *Good Things* Jane Grigson gives, in my opinion, the best recipe of all, when she adds curry powder at the beginning of cooking. Because of its sweetness, I think parsnip needs spices to offset and counter the sugar. All the other recipes that mix it with leeks, for example, in a kind of variation of Vichyssoise, are too sweet for my taste. Instead of curry powder, you could use ginger, chilli and lime, which works very well – as in the recipe for parsnip moulds you will find overleaf.

PARSNIP CAKES

Boil 3 or 4 parsnips in salted water until quite soft. When cool, mash them with a beaten egg and a few breadcrumbs, season with pepper and salt. Make into small flat cakes, roll in egg and breadcrumbs, and fry till a golden brown. Sufficient for 4 or 5 people.

PARSNIP FRITTERS (1)

Serve these with chops and cutlets. First boil in salted water till tender, then cut into rings 1 cm / ½ in thick. Dip in frying batter, fry in deep fat, scatter salt over and serve very hot.

PARSNIP FRITTERS (2)

Cut some parsnips into small pieces, then boil until they are soft. Mash them with a fork, then add a little chopped parsley, onion and any scraps of cold meat, also a few drops of relish. Mix all together and form into fritters, roll in breadcrumbs and boil in fat until they are a golden brown.

GLAZED PARSNIPS WITH MARMALADE

675 g / 1½ lb small parsnips, trimmed
30 g / 1 oz butter
1 heaped tbsp caster sugar
1 heaped tbsp bitter marmalade
sea salt and freshly ground black pepper

Quarter and slice the parsnips into 8 cm / 3 in chunks (no need to peel). Plunge them into boiling salted water for 3 minutes. Drain well.

RIGHT *clockwise from left: Parsnip fritters, carrot and red pepper terrine (page 216), parsnip cakes*

Melt the butter and the sugar in an oven dish and place the parsnips in this, turning them over so that they are well covered. Season and add the marmalade. Place the dish in an oven preheated to 190°C/375°F/gas5 for 20–30 minutes, or until they have browned and look well glazed. Excellent with game or ham.

PARSNIP MOULDS

450 g / 1 lb parsnips, peeled, trimmed and cut into chunks
30 g / 1 oz butter
55 g / 2 oz root ginger, peeled and grated
1 green or red chilli, chopped small
1 egg
250 ml / 8 fl oz double cream
sea salt and freshly ground black pepper

Boil the parsnips in a little salted water for 5 minutes, then drain well and reserve.

Melt the butter, add the ginger and chilli and cook over a low heat for a few minutes.

Blend the parsnip, butter, spices, egg, cream and seasoning together and pour into individual ramekins. Cook in an oven preheated to 190°C/375°F/gas5 for 20–30 minutes, or until the centre has risen.

Leave to cool, then unmould and serve with a tomato or red pepper coulis and a few salad leaves.

PARSLEY

Petroselinum crispum

Where would the cuisines of the world be without this ubiquitous herb? Not only because it gives its name to the family and is – with its gastronomic companion, garlic – among the very oldest flavourings stemming from the Mediterranean area, but also because its general use in cooking makes necessary its appearance in these pages.

Parsley and rue often bordered Greek gardens. The victors in the Isthmian games would be crowned with chaplets of parsley, possibly because it was said that Hercules after killing the Nemean lion (he choked it in his arms) crowned himself with the herb. In battle, the warriors in Homer fed their horses with it, while wreaths would also be made from it to be laid upon the tombs of the dead. All this reverence and esteem clung to the herb so that by the seventeenth century Culpeper thinks parsley very comfortable to the stomach and good for wind and Aubrey (aware of its ancient symbolism) mentions that only the tops of the leaves may be admitted to the table, for it is more proper to be used in forcemeat stuffings for fowl. He approves of it in medicinal drops, however. The ancients were rather more enthusiastic than Aubrey gives them credit for. Dioscorides gave parsley its name, Pliny said that sauces and salads should never be without it, Horace decorates his dining room with roses and parsley.

The Romans knew five different kinds of parsley and so do we, but possibly they are not still the same kinds. The most familiar to us is the curly kind or moss-curled parsley (this is the one to use for deep-frying in whole sprigs, see below). It grows better in the northern climes than in the Mediterranean, as it will survive colder temperatures and lack of sun. It is a great pity that this type has been demoted to the role of garnish and is often left ignominiously at the side of the plate uneaten.

The least known variety is Neapolitan parsley, which is grown as much for its leaf stalks as it is for its leaf. Also called celery-leaf parsley, it grows in southern Italy and was little known outside this area until this century. The stalks of the plant are hollow and they are blanched and eaten like celery. Pliny refers to a Macedonian parsley or black parsley where the stalks could be eaten, and it is thought that the Neapolitan parsley could well be the same one. The other three parsleys look a little similar: the plain-leaf, the fern-leaf and the turnip-rooted or Hamburg parsley. The flavours of all five types are broadly similar, but with slight differences – yet not different enough to suggest one type against another should be used in any particular dish.

HAMBURG PARSLEY

Petroselinum crispum var. *tuberosum*

Hamburg Parsley had a great vogue in the eighteenth century, when it would be grown for its root. Miller in his *Gardener's Dictionary* of 1771 says: 'This is now pretty commonly sold in the London markets, the roots being six times as large as the common parsley. This sort was many years cultivated in Holland before the English gardeners could be prevailed upon to sow it. I brought the seeds of it from thence in 1727, but they refused to accept it, so that I cultivated it several years before it was known in the markets.'

Now the only chance of eating it is to grow it in your own garden. Once dug up the root looks very like parsnip, but tastes more like celeriac, yet sweeter and more aromatic.

If you are lucky enough to grow Hamburg parsley in your garden, then clean the root and treat it like parsnip. Chopped into chunks and boiled in a little water with a pinch of salt, it will cook within about 4 minutes.

MEDICINAL

The leaves of parsley are dried to make parsley tea and the seeds of the plant are used for the extraction of its oil, Apiol. Parsley has a carminative, tonic and aperient action; it is also a diuretic; and a poultice of the fresh leaves have been used to relieve stings and bites from poisonous insects. The chewing of parsley leaves is suggested to cleanse the breath and keep the skin healthy.

NUTRITION

The leaves are tremendously high in carotene and potassium, calcium and vitamin C. They also have small amounts of vitamin E, riboflavin and thiamin.

CHOOSING AND STORING

Parsley has a short shelf life and it is best to use it quickly after purchasing or picking. Plunge the stalks in a glass of water, keep it in a shady part of the kitchen and use within three days. Parsley sold in airtight plastic bags seems to keep fairly well in the salad drawer of the refrigerator, but not for more than two days. Even then you might still find a soggy strand lurking among the foliage. Parsley is also sold chopped and frozen and this is an excellent standby to keep in the freezer, if you intend to use parsley for a sauce or to add to a hot dish. This is a good way of freezing your own glut of parsley if you are unable to keep it growing through the winter.

COOKING

We tend to think of cooking with parsley only in terms of chopping it and using it in sauces or adding it to stews or vegetable dishes. To my mind, for example, broad beans in parsley sauce is one of the combinations made in heaven. But sprigs of curled parsley can be deep-fried. DEEP-FRIED PARSLEY may sound unlikely, but try it – the result, I promise, will astonish you. The sprigs need only about 5 seconds in hot fat, when they will turn dark green and frazzly. Take them out and blot on absorbent paper, then eat at once. They are so good that they are best served as appetizers. Take this idea one step further by dipping the sprigs into a light and seasoned tempura batter before deep-frying: they will need a little longer to cook, say 12 seconds. You can serve these with other tempura vegetables, if you like, or as an accompaniment with fish. The more densely curled the parsley is, the better it is for cooking. Never try to cook flat parsley or Hamburg parsley – it will just go limp, then shrink to a blackened speck.

When using chopped parsley in a sauce, make the sauce first and add the parsley only at the last minute so that it simply heats through. Parsley must never be cooked, just heated – except when fried as above.

ROCK SAMPHIRE

Crithmum maritimum

This plant much loved in antiquity and highly popular until the eighteenth century can be found only in the wild. It grows on rocky coastlines from the Crimea to Cornwall and Ireland; certainly it is still flourishing all around the coasts of Britain and France. The samphire which you can now buy from fishmongers is marsh samphire and a completely different plant (see page 58). The popularity of rock samphire (sometimes called sea fennel) was so great that demand outstripped supply in the nineteenth century, so marsh and golden samphire were substituted for it in the London markets. Customers found this both annoying and disappointing, hence both samphires declined in popularity. Also, as the urban sprawl continued to encroach upon the countryside, people turned towards cultivation in their back gardens and allotments and forgot about gathering wild foods from the countryside. It takes only one generation for such knowledge, passed on for thousands of years, to be forgotten forever.

Golden samphire (*Inula crithmoides*) also grows in salt marshes and on sea cliffs, but it is rare, plentiful only on the Isle of Sheppey, which is conveniently near for the London markets. The young branches of golden samphire were mixed in with rock samphire. Green writes in deep resentment in his *Universal Herbal* (1832): 'It is a villainous imposition because this plant has none of the warm aromatic taste of the true samphire.'

Pliny wrote that Theseus had a meal with samphire before leaving to fight the Minotaur. Both the Greeks and the Romans used samphire in salads, sometimes lightly steamed, and they also took it in wine to clear the complexion and give a happy expression. The name derives from Saint Peter, though the French originally called it '*perce-pierre*' – 'rock piercer'.

It has, according to Gerard, 'many fat and thicke leaves somewhat like those of the lesser Purslane, of a spicie taste, with a certain softnesse.' Lady Fettiplace in her book (1604) gives a recipe for pickling samphire with other pickling recipes for items as various as purslane and broom buds.

It was John Evelyn's favourite vegetable. He talks of its ability to sharpen the appetite, preferable to other herbs and salads. Evelyn cultivated it, getting his seeds from France. He observed that the cultivated kind did not pickle so well, as it had a more tender leaf and stalk, but for salads there was nothing better. Here is Evelyn's recipe for pickling samphire:

'Let it be gathered about Michaelmas or in the spring and put two or three hours into a brine of water and salt, then into a clean tinned brass pot with three parts of strong white wine vinegar and one part of water and salt or as much as will cover the sampier, keeping the vapour from issuing out by pushing down the pot lid, and so hang it over the fire for half an hour only. Being taken off let it remain cover'd till it be cold and then put it up into small barrels or jars with the liquor and some fresh vinegar, water and salt, and thus it will keep very green. If you be near the sea that water will supply the place of brine. This is the Dover Receit.'

What kept the colour bright green was the copper salts (certainly toxic) produced by the vinegar reacting with the brass of the pot.

Shakespeare mentions 'half way down/Hangs ore that gathers samphire, dreadful trade.' Dorothy Hartley confesses to being puzzled by this quotation, for Shakespeare – being sound on food – would be bound to know that samphire also grows on shingle. But gathering samphire from the face of steep cliffs was well-known. Robert Turner in 1664 wrote about samphire gatherers on the cliffs of the Isle of Wight, where it is incredibly dangerous to gather, 'yet many adventure it, though they buy their sauce with the price of their lives'. The Lord of the Manor of Freshwater charge a yearly rent from the cliff gatherers who would find gulls' eggs and samphire on the green shelves in the 600-foot cliffs. The Islanders made a sauce out of samphire and butter; much of it they pickled for themselves, the rest they sent to London. But it was not the only island to export their samphire. Thomas Cogan in *The Haven of Health* (1584) says that the Isle of Man also pickled samphire in casks of brine. In the nineteenth century wholesalers would pay up to four shillings a bushel for it.

Though scarcity in both coastal and inland areas caused the popularity of rock samphire to decline at the end of the nineteenth century, not everyone forgot it. Dorothy Hartley writes engagingly as ever of the effect of samphire upon some people:

'Samphire grows on rough shingle; there is a lot on the pebble ridge at Bideford. You can smell it before you find it. Among all the delicate subtle scents of the country, samphire holds unique place. People who dislike it say it smells of sulphur, but others sniff it ecstatically, and seem to make themselves slightly drunk on the aroma. Then, surprisingly, someone who dislikes it the first time will try again, and find they like it extremely well! It's the most complete puzzle – I have never yet met anyone who was neutral to it. There is something "magic" about samphire.

'You can sometimes buy it in country markets, but the liking for it is so uncontrolled that you find some families – miles away from the sea – getting it sent to them, as a delicacy, while the rest of the community look on in bewilderment. Try it; for if you like it you will have added a very pungent, enjoyable, and health-giving item to your diet.'

She goes on to quote a recipe of 1650 for Samphire Hash. This was pickled cucumbers, capers, samphire, lemon, pepper and nutmeg boiled in strong stock with a little vinegar and thickened with egg yolk. It was used as a sauce for mutton and was garnished with more samphire and barberries.

It was a great thrill to find it growing one Christmas on the Mediterranean island of Gozo. Samphire flourishes on those high rocky cliffs; great shrubs of it go unnoticed and unharvested by tourists and Gozitans alike. Recently I found it growing in profusion all over the rocks at Antibes, again entirely ignored by both tourists (there, one supposes, to have their palates stimulated) and French gourmets. I picked a huge bunch and later that evening we had a feast. How sad it is that we have forgotten how to feed ourselves from the wild.

ABOVE *Rock samphire growing on a cliff face*

Samphire flourishes all through the summer months. If you spot it on isolated beache or rocky coast (the *Reader's Digest Field Guide to the Wild Flowers of Britain* has the best illustration), pick the leaves and steam for 5 minutes or poach in boiling water for three, then serve with a garlicky vinaigrette or a lemon butter sauce.

TROPICAL
AND
EXOTIC
VEGETABLES

Gramineae

SWEETCORN *Zea mays*

Malvaceae

OKRA *Hibiscus esculentus*

Convolvulaceae

SWEET POTATO *Ipomoea batatas*

Araceae

TARO *Colocasia esculenta*

Dioscoreaceae

YAM *Dioscorea* spp.

Lauraceae

AVOCADO *Persea americana*

The miscellany of vegetables in this chapter do not belong to one family but are the single representatives of a series of different families. They are native to tropical, subtropical or warm temperate regions. Almost all of them are exotic in the sense that they are eaten – if not cultivated – in parts of the world far from where they originate. Two of them – okra and sweetcorn – are grown by more adventurous gardeners, but on the whole what these vegetables have in common is that they are imported to Western markets and bought in by cooks as they are needed. Most used to be available only in specialist ethnic stores, but increasingly they are found in supermarkets.

THE GRASS FAMILY

Gramineae

Grasses were the first food plants to be cultivated by human kind, probably by accident. Wild grasses flourish on poor soil, among stones or in gravel; they like a dry and a wet season (humidity is too lush for them), they need to germinate quickly in spring rain and to have their life cycle concluded with mature seeds ready to fall before the ground thoroughly dries in the height of summer. They, therefore, colonized the bare ground and rubbish heaps provided by early peoples. Because they had evolved large food reserves in their seeds, these were found to be good to eat.

Economically they are the most important group of flowering plants because of these nutritious grains. Grasses also provide food for all herbivorous animals, including the domesticated ones, and shelter for many other creatures that were eaten. Sugar cane has been grown since the earliest times and is still a highly important economic crop due to the world's insatiable desire for sweetness. There are, among the Gramineae, two vegetables, one of which since 1492 has assumed a major part in the world economy – maize or sweetcorn. The other is bamboo which I have omitted from this book as it is still only really available in tinned form in the West.

SWEETCORN

Zea mays

The first record of maize is in the sacred book of the Quicke Indians of Guatemala in the eighth century. Here the story is told of the discovery of white and yellow maize which the gods ate so that man became strong. Here, too, is the first reference to maize being the Indians' first mother and father, the source of life. It was so much esteemed by the Indians that in the palace gardens of the Incas there were decorations of gold and silver maize with all the grains, stalks, spikes and leaves depicted in fine detail.

Columbus first saw the corn in Cuba in November 1492, noting 'a kind of grain called *maiz*, of which was made a very well-tasted flour.' The early explorers soon discovered that corn came in various colours, red, white, yellow, blue and even black, but blue corn was the most valued. They were also impressed by the Indians' agriculture. Diego Columbus once estimated that a corn, bean and squash plantation he walked through was 18 miles long. In Peru it was noticed that in each hole three grains of maize were sown, along with a fish head to provide slow-release fertilizer. Thomas Hariot, reporting on the first English settlement in Virginia, talks of corn as a grain of 'marvellous great increase'. He observed that the corn was sown early, with three dead fish laid over the heaped up mound; beans were planted later and would twine around the corn stalk. The first settlers learned not only how to grow corn but when to pick it and how to cook it – discovering how much more delicious corn is when picked and cooked immediately, before the sugar turns to starch. They learned how to cook corn, beans and peas together 'by boyling them all to pieces into a broth', and also noted that 'sometimes they bruse or pound them in a mortar and thereof make loaves or lumps of dowishe bread'. Sometimes this bread contained dried huckleberries. It is interesting to note how much traditional American cooking owes to the Indians, even the names: succotash (*misickquatash*), pone bread and blueberry muffins, hominy or grits (*rocka hominy*), samp (*nasaump*) – a kind of meal pottage made of unparched corn, which for years the settlers ate as both breakfast and supper with milk and butter added. It was also called hasty pudding. The corn mush could be left to cool in a loaf tin, then sliced, dipped in egg yolk and breadcrumbs and fried in bacon fat.

Sturtevant says, 'the culture of corn was general in the New World at the time of the discovery; it reigned from Brazil to Canada, from Chile to California; it was grown extensively in fields; and it had produced many varieties – always an indication of antiquity of culture. It furnished food in its grain, and, from its stalks, sugar to the Peruvians, honey to the Mexicans and a kind of wine or beer to all the natives of the tropics.'

It was a different story in Europe. Brought to Spain by Columbus, maize took some time – until 1610 – to move to Sicily and from there to the rest of Italy. However, by 1650 Italy had taken to it in a big way. They, too, made a porridge out of the grains (they had originally used millet, spelt or chick pea flour). They, too, let the porridge cool before slicing and frying or grilling it; in northern Italy *polenta* is still eaten with enthusiasm.

But where the poor lived off maize and nothing else, they inevitably became ill. The symptoms of *mal de rosa*, as Philip V's doctor, Gaspar Casal, called it, included a reddening and roughness of the skin across the arms and face, accompanied by general lassitude, headaches, diarrhoea and insanity. The poor of Andalusia first showed the symptoms, then northern Italy, followed by France, Hungary and Romania. It was noticed that the spread of pellagra occurred wherever maize was cultivated. If maize was only part of the diet, with vegetables, eggs and fish, health was retained. It was where maize comprised all of the diet that pellagra inevitably followed. Yet millions of people in Mexico and the southern States of America ate little else and pellagra was unknown there – that is, up to the beginning of this century. By 1912 there were thousands of cases of pellagra in the Southern cotton-growing areas. What was different in Mexico or in the early 1900s in the American South?

First, the Indians would eat their corn with beans, squash and a pinch of wood ash in the pot. The last, it was believed, softened the skins of the corn kernels and made them easier to grind and digest. Ash in lime, or alkali as we now know it, releases the niacin and lysine in the corn which would otherwise have been unavailable to humans. Also, beans and squash complement corn nutritionally and turn it into a staple food.

ABOVE *Maize for polenta*

In 1905 a cereal mill called the Beall Degerminator started producing factory-refined corn. It removed the germ which contained the oil so that the mush or porridge that the poor now ate was nutritionally inadequate. It took many years for this to be discovered; the story is in *The Food Factor* by Barbara Griggs.

If you are intrigued by the story of maize, Margaret Visser is fascinating in *Much Depends on Dinner*. Here you will learn that you cannot buy anything in the North American supermarket which has been untouched by maize. Meat and milk are largely corn, because livestock and poultry are fattened on cornstalks and corn; frozen meat and fish have a cornstarch coating, the golden colouring of soft drinks and puddings is corn and so on, while corn oil, corn syrup and cornstarch all permeate other foods.

NUTRITION

The fresher the corn is, the more nutritious; it is then a good supply of vitamins A, B and C, and is high in phosphorus, potassium and sulphur. It is a good source of starch in the diet, but the protein is of lower nutritional value than that of other cereals. The yellower the maize, the more carotene it contains.

VARIETIES

Five different types of maize or Indian corn known to the native Americans are still grown today. Pop and flint corn have high protein content and a hard, waxy starch. Dent corn, with a soft starch that produces a dent in the kernel, is grown for animal feed. Flour corn is grown now by the Indians only for their own use; it is low in protein and has a waxy starch. Both flour and flint corn have variegated kernels.

Sweetcorn, the fifth type, is a variety that stores more sugar than starch, which is why it tastes so good eaten fresh from the cob without the processing that the coarser maizes require.

Native to subtropical Central America, maize needs just the right combination of moisture and warmth to swell and ripen the grain, but breeders have produced a range of early-maturing sweetcorn varieties that succeed reasonably well in sunny, sheltered gardens in more temperate climes.

CHOOSING

If you succeed in growing sweetcorn, there is nothing like the picking and cooking of the first ripe cob. The sweetness of the kernels sings on the palate and you know that Mark Twain is right when he says a cauldron of boiling water should be set up in the midst of the maize field so that the corn can be thrown straight in as it is picked. (Butter, preferably unpasteurized, is the only accompaniment to such a feast, with a little sea salt and a hefty grind of the pepper mill.)

To tell if a cob is ripe, keep a sharp look out on its size. Watch the silky threads, too, for signs, of turning from light gold to brown. When a cob has

BELOW Polenta served with pigeon breasts in a redcurrant sauce

grown 18–20 cm / 7–8 in long and feels full and slightly bumpy, gently part the outside husky covering and see whether the kernels are round and fat. If they look golden, glossy and give a little when pressed, they are ripe. If small and creamy in colour, leave the cob to grow. Another test is to pierce a kernel with a fingernail: pick the cob if a milky juice appears.

If the silk is black when cobs have reached the shops, do not worry; the kernel will be less sweet (as the sugar will have turned completely to starch), yet none the less perfectly pleasant and satisfying. The corn will keep for a few days, but it is much better to use it at once.

In the last few years baby sweetcorn from Thailand has been available. This comes in a cling film package and will keep happily for a week in the refrigerator. But again, all vegetables are best used as soon as possible, and packaged ones are no exception.

COOKING

Sweetcorn needs to have its husks and silken threads stripped off, then slice away the base and boil for 6–10 minutes. Do not use salt in the water and test the kernels by sticking the point of a knife in after the 6 minutes. When cooked, drain well in a colander and serve with melted butter, sea salt and freshly ground black pepper. I always remember the delight of my small son when, aged five, he discovered that sweetcorn was eaten in a gorgeously messy fashion – picked up and gnawed, while butter became smeared over fingers and cheeks. This, of course, is partly why it is such a popular and satisfying food. It is eating at the trough and socially permissible. Of course, it was not always so. *Hints on Etiquette* published in 1844 in America lays down the law: 'It is not elegant to gnaw Indian corn. The kernels should be scored with a knife, scraped off into the plate, and then eaten with a fork. Ladies should be particularly careful how they manage so ticklish a dainty, lest the exhibition rub off a little desirable romance.'

In many parts of the world sweetcorn is street food, and the cobs are roasted over charcoal. Certainly this is one excellent method of cooking it in the summer when the barbecue is alight. Cobs will cook quietly at the side when other foods occupy the centre. They need about 10 minutes' cooking over high heat, but if pushed to one side they can keep warm or cook very slowly for up to an hour.

You can also roll the cobs in a flavoured olive oil and grill them, turning to cook on all sides, but they need a watchful eye for they can easily burn. They can also be wrapped in foil and baked in an oven at 190°C/375°F/gas5 for 20–25 minutes.

Once cooked, the kernels can be sliced from the cob and used for other dishes. This might seem a chore, but it takes very little time. However, you might be tempted to use tinned or frozen corn. If so, do make sure there is no added sugar in these products, which make the kernels unbearably sweet.

Baby corn is perfect for many stir-fry dishes, used whole, sliced in two lengthwise, or cut in smaller pieces across. These have a delicate flavour and if you want to cook them plainly boiled, they only need 1 minutes' cooking before being ready.

CORN PUDDING

This recipe from the *American Heritage Cookbook* has been handed down in the family of General Daniel Morgan, a Revolutionary War hero. Try it as an alternative to Yorkshire pudding.

85 g / 3 oz plain flour
3 eggs, beaten
30 g / 1 oz butter, melted
300 ml / ½ pt single cream
1 tsp sea salt
corn kernels cut from 3 cooked cobs, or 170 g / 6 oz defrosted frozen or drained canned kernels

Mix the flour with the eggs to a smooth paste, then add the butter, cream and salt. Beat vigorously and finally add the corn. Let the batter stand for an hour. Meanwhile, preheat the oven to 190°C/375°F/gas5. Pour the batter into a buttered oven dish, place in a pan of hot water and bake for 30–40 minutes, or until the pudding has puffed up and is golden. A knife inserted in the middle should come out clean, but as with soufflés I rather prefer the centre to be moist and runny.

CORN OYSTERS

For 10 – 12

corn kernels cut from 2 cooked cobs, or 115 g / 4 oz defrosted frozen or drained canned kernels
30 g / 1 oz maize flour
30 g / 1 oz plain flour
2 eggs, beaten
½ tsp sea salt
freshly ground black pepper
sunflower oil for frying

In a bowl, mix the kernels with the flours and eggs, then add the seasoning.

Heat the oil in a pan and fry spoonfuls of the mixture until brown on both sides.

FRIED BABY SWEETCORN WITH OKRA AND GARLIC

2 tbsp olive oil for frying
1 pack baby sweetcorn, sliced in half lengthwise
225 g / 8 oz okra, sliced lengthwise
5 garlic cloves, sliced
1 tbsp chopped coriander leaves
½ lemon
sea salt

Heat the oil in a pan and throw in the sweetcorn, okra and garlic. Fry briskly for 3 minutes or so, until the vegetables are tinged with brown. Serve on a platter sprinkled with the coriander, lemon and salt.

SUCCOTASH

115 g / 4 oz haricot beans, soaked and boiled until tender
corn kernels cut from 2 cooked cobs, or 115 g / 4 oz defrosted frozen or drained canned kernels
30 g / 1 oz butter
2 tbsp double cream
generous handful of chopped parsley and chives
sea salt and freshly ground black pepper

Drain the haricot beans and add the corn kernels, butter and double cream. Simmer for a few minutes, then season and stir in the chopped fresh herbs. Serve the succotash at once.

POLENTA

1 tsp sea salt
140 g / 5 oz cornmeal
55 g / 2 oz Parmesan cheese, grated
30 g / 1 oz butter

Bring 575 ml / 1 pt water and the salt to the boil, add the cornmeal to another 300 ml / ½ pt of cold water in a bowl and combine, stirring all the time. Once it has boiled again, lower the heat and simmer for 10 minutes.

Pour into a buttered loaf tin, let cool and refrigerate for an hour or two.

Cut into 1 cm / ½ in slices and place in a shallow baking dish. Sprinkle with the Parmesan, dot with the butter and grill until golden.

THE MALLOW FAMILY

Malvaceae

This comprises a large group of flowering plants, herbs, shrubs and trees. They include such familiar flowers as hollyhocks, hibiscus and abutilon. Economically cotton is the most important plant, but a close second in this century has to be the nuts which are an ingredient of Coca-Cola. Cacao nuts yield cocoa butter for chocolate, while kola nuts, from *Cola nitida*, are high in caffeine and the glucoside kolanin which is used extensively for flavouring drinks. The one vegetable of the mallow family is now a fairly common sight to us – okra.

OKRA

Hibiscus esculentus

Okra is also called 'gumbo' or 'ladies' fingers'. The multisided green seed pods from this annual plant may be long and slender – hence the name ladies' fingers – or are sometimes shorter and more rounded in shape. Though we might associate this vegetable immediately with the Southern States of the USA, it originated in tropical Africa. The name okra comes from the Twi language of the Gold Coast, now Ghana, which called okra '*nkurama*', while gumbo comes from the Angolan word, '*ngombo*' or '*kingombo*', which became attached to the dish that okra was an ingredient in.

Though the evidence for its existence in Ancient Egypt is slight, it was almost certainly there. A fruit called '*banu*' which is fairly close to the Arabic *bamia* is mentioned in a papyrus. Okra is now extensively cultivated in Egypt, elsewhere in Africa and in India. From Africa it was also taken west to Brazil some time before 1658. The Spanish Moors knew okra well; we have an account of a resident in Seville who visited Egypt in 1216 and describes okra as being eaten when young and tender.

The most striking quality in okra is its mucilaginous seeds. If the vegetable is cooked with liquid, the contents ooze out into a gummy thickening. In the Southern States of the USA this was used in soups and stews in a dish adapted from the Indians which used to be thickened with filé powder (the dried pounded leaves of the sassafras tree), also mucilaginous. Unlike okra itself, the filé powder could not be cooked, even simmered, or else it made the gumbo stringy. Gumbos could be made from various mixtures, including fish, shellfish, frogs' legs, turtle meat, chicken, or even ham, shrimps, tomatoes and oysters; it is, in fact, not unlike the original ingredients of the *paella* or *potage bonne femme*. Whatever was at hand was turned into a one-pot meal. A visitor to New Orleans in 1805 spoke of the quantity of shrimps eaten and added, 'also a dish called gumbo. This last is made of every eatable substance and especially of those shrimps which can be caught at any time.'

But there was another gumbo – Gumbo Z'herbes – which originated in the Congo and was introduced to New Orleans by the black slaves. This was made

BELOW *Okra at market*

with any green leaves, herbs and seasonings at hand; in Louisiana, the ingredients changed to the herbs and greens sold there in the French market by Cherokee and Choctaw Indians. The Gumbo Z'herbes was traditionally served on Maundy Thursday.

NUTRITION

Okra contains generous amounts of calcium, magnesium, potassium and phosphate; it is high in vitamin C and carotene, with traces of thiamin, riboflavin and vitamin B6. Few of the nutrients are lost when okra is fried briskly in olive oil; more are destroyed by slower simmering in a gumbo.

CHOOSING

In temperate climates you can grow varieties of okra under glass or in favoured warm spots. Pick the pods when they are 5 – 8 cm / 2 – 3 in long.

Okra is cultivated in large quantities in the southern States of the USA for the canning industry. Pods are picked every few days to ensure that only the smaller ones go on sale: older ones tend to be tough and fibrous. I would not recommend using canned okra; it is already cooked, so could at a pinch be added to stews at the last moment. (The okra or *bhindi* curries you find in many Indian restaurants often look as if they have come out of a tin and just been spiced and sauced in the kitchen.)

Fresh okra is another treat altogether. Choose small pods that are bright green and firm; they should be just springy when gently squeezed. Once they begin to soften, or show any brown patches, they are too old and should be dismissed. If you have to keep them for a few days, put them in the cool drawer of a refrigerator. Really fresh okra will keep there for up to a week.

PREPARING AND COOKING

Rinse okra under a tap and drain it well; you do not have to prepare it in any other way. Most cookery writers tell you to snip off the stem – there is no need. It is perfectly edible. If you do, there is a chance you may snip into the top and expose the

RIGHT *Seafood-okra gumbo (overleaf)*

seeds and pith. This *can* be a mistake, but it depends totally on how you are going to cook it, and whether you need the mucilaginous quality or not. On the whole, I am against vegetables that make a slimy jelly, so I am not mad about gumbo. But I do love okra.

If you want to try the taste of okra plain, which is of high quality, simply throw into a hot pan with a little olive oil (and a chopped clove of garlic, if you wish) and fry for a few minutes. There is nothing mucilaginous about okra cooked in this way. If you add a chopped chilli to the oil, this makes an excellent dish of appetizers for handing around to friends with the drinks. Frying also allows you to gauge whether you want the okra fairly soft or crisp. Just cook a little longer if you want the latter, which is what I prefer. You might think that it is exposing the pods, coupled with heat, that makes the gummy juices extrude. But no, for you can cut the okra in half down the centre and fry both halves; this exposes the seeds and makes the okra look very pretty once it is served.

The first time I cooked okra was in the 1950s when there was very little literature about it and none I had seen. I treated it like any other vegetable, throwing the pods into a little salted water and poaching them for about 5 minutes. At the end of this time I looked into the pan and found something fit for a horror film, a kind of mess of slime, as if rotting vegetation and frog spawn had fused. I threw it away and never tried okra again for many years. I missed a lot.

Okra needs only one or two minutes' frying. The mucilage will only seep out if you add liquid and cook longer than a few minutes. For most dishes you do not need longer.

OKRA WITH RED PEPPERS

2 tbsp olive oil
1 red chilli, chopped
3 garlic cloves, sliced
1 red pepper, deseeded, cored and sliced
1 orange pepper, deseeded, cored and sliced
2 or 3 tomatoes, skinned and chopped
450 g / 1 lb okra
sea salt and freshly ground black pepper

Heat the olive oil and throw in the chilli and garlic, cook for a moment, then add the peppers. Fry them for about 5 minutes, then add the tomatoes and the okra. Cook for another 3-4 minutes. Then add the seasoning and serve.

SEAFOOD-OKRA GUMBO

for 6 to 8

I have included one gumbo dish (from the *American Heritage Cookbook*) because they are so closely identified with the vegetable.

450 g / 1 lb raw prawns in the shell
55 g / 2 oz butter
450 g / 1 lb okra, sliced
2 onions, finely chopped
1½ tbsp flour
225 g / 8 oz tomatoes
12 oysters in their liquor
2 tsp sea salt
1 garlic clove, crushed
pinch of cayenne, or ¼ pod red pepper
Tabasco
Worcestershire sauce
225 g / 8 oz crab meat
boiled rice

Shell the prawns and sauté in half the butter for several minutes or until they turn a bright coral colour.

Heat the remaining butter in a large saucepan, add the okra and cook, stirring frequently, until tender. Stir in the onion and cook for several minutes, then stir in the flour until smooth. Add the tomatoes and cook the mixture for several minutes longer.

Add enough water to the oyster liquid to make 2 quarts of liquid. Stir this into the okra mixture and add salt, garlic and the cayenne or red pepper pod. Simmer for 1 hour. Add the prawns and simmer for another 5 minutes.

A few minutes before serving, add the oysters and their liquid and cook over a low heat until the edges begin to curl. Then add Tabasco and Worcestershire sauce to taste, add the crab meat and heat through. Serve the gumbo in soup plates over boiled rice.

OKRA WITH ONIONS AND EGGS

This is a Parsee dish I have adapted from one of Madhur Jaffrey's recipes.

3–4 garlic cloves, chopped
2–3 onions, sliced
1 green chilli, finely chopped
(pith and seeds removed for a milder taste)
450 g / 1 lb okra
1 tsp turmeric
2 tbsp mustard oil
3–4 eggs, beaten
sea salt and freshly ground black pepper

Heat the oil and throw in the garlic, onions, chilli and okra, sprinkle over the turmeric and cook, stirring frequently, for about 4 minutes.

Add the eggs and seasoning and let them set over the vegetables. An excellent supper dish or snack.

THE MORNING GLORY FAMILY

Convolvulaceae

This family of flowering plants includes 1,400 species widely cultivated for their colourful funnel-shaped flowers. In my garden in Greece the blue morning glories had an intensity of colour which they seem never to have elsewhere. Plants of the family tend to be twining climbers, woody vines and herbs. Gardeners are all too aware of bindweed, which races to stifle your favourite seedlings. The family includes one important food plant – the sweet potato.

SWEET POTATO

Ipomoea batatas

This is the potato that first came to Europe. Columbus encountered it in Haiti in 1492, and in 1514 Peter Martyr mentions *batatas* as being cultivated in the Honduras, giving the names of nine varieties. For years there has been confusion as to which potato was brought to Europe from Virginia and which potato Raleigh planted in his gardens at Youghal in southern Ireland. In the portrait from the frontispiece of his *Herball* (1597) John Gerard holds the leaf and flower of the potato plant: the flower is unmistakably of the morning glory type.

By 1526, Oviedo tells us, sweet potatoes have been carried to Avila in Castile, where the Spaniards took to them with much enthusiasm. They came in three different kinds – red, yellow and white. Within a few years they were in general cultivation in Spain, Portugal and probably also Italy. They had also been taken from the Americas to China and the Philippines, where they were cultivated.

The sweet potato had been grown in the warmer parts of the Americas for centuries and had become a staple plant. We do not know for how long, because details of neolithic cultivation are inevitably lost to us, but the sweet potato had certainly crossed the Pacific Ocean and become established in Tahiti, the Fiji islands and New Zealand. Unlike the coconut or the bottle gourd, the sweet potato cannot float for a long time and so be dispersed naturally by the oceans; it had to have been taken there. It might have got to Polynesia when Indians in frail crafts fled the Spanish invaders, or it could have come earlier from some Indian armada of explorers, sliced and dried like taro as part of their provisions.

It is surely due to the sweet tooth of the Renaissance elite that the sweet potato became an instant success; unabashed, they added more sugar, spices and dried fruit to the vegetable when they cooked it, as in the pies made for Renaissance princes. Richard Hakluyt ate some sweet potatoes in 1589 and writes: 'These potatoes be the most delicate rootes that may be eaten, and doe farre exceed our parseneps or carets.'

Gerard writes, 'The Potato roots are among the Spaniards, Italians, Indians [he was still under the impression that Columbus had discovered India], and many other nations, ordinaire and common meat; which no doubt are of mighty and nourishing parts, and doe strengthen and comfort nature; whose nutriment is as it were a mean between flesh and fruit, but somewhat windie; yet being rosted in the embers they lose much of their windinesse, especially being eaten sopped in wine.

'Of these roots may be made conserves no lesse toothsome, wholesome, and dainty, than of the flesh of Quinces; and likewise those comfortable and delicate meats called in shops, Morselli, Placentulae, and divers other such like.

'These roots may serve as a ground or foundation whereon the cunning Confectioner or Sugar-Baker may worke and frame many comfortable delicat Conserves and restorative sweet-meats.

'They are used to be eaten rosted in the ashes. Some when they be so rosted infuse and sop them in wine: and others to give them the greater prace in eating, do boile them with prunes and so eat them: likewise others dresse them (being first rosted) with oile, vinegar, and salt, every man according to his owne taste and liking. Notwithstanding howsoever they be dressed, they comfort, nourish, and strengthen the body.'

Because Gerard distinguished the white potato from the sweet as the potato from Virginia, historians and botanists were confused for some time. The white potato from Peru failed to find acceptance in Europe for over 200 years, unlike the sweet. Marnette's *The Perfect Cook* includes a recipe for potato pie in 1656, obviously for the sweet potato as it includes cinnamon, nutmeg, mace, grapes and dates. It took another hundred years for the white potato to be named; in the *Oxford English Dictionary* of 1775 'sweet potato' makes its first appearance. To add to the confusion, the sweet potato in the United States and the Caribbean is often called a yam, because it reminded the slaves of the vegetable they had known in Africa.

The sweet potato has an unusually high yield, four times that of rice. It tolerates poor soils and is resistant to drought, making it an important secondary crop throughout subtropical countries. However, even in the warmer climate of southern Europe it did not catch on, any more than the white potato had done. Today in Europe the sweet potato is still bought only by those peoples who have emigrated there from countries where it is an established part of the diet. What a lot most of us are missing.

NUTRITION

Sweet potatoes show markedly antioxidant properties, they lower blood cholesterol and oppose the free radicals which can ravage body cells. The darker the orange of the potato, the greater the concentration of beta-carotene. The vegetable is also rich in vitamins C and E.

VARIETIES

There are hundreds of varieties spread around the world. They come in ovoid or tapering root shapes, the skin can be smooth or ribbed, pink, red, purple, brown or yellow. The flesh too comes in a variety of shades from white to cream, golden and sometimes almost a deep tangerine – the colour of a pumpkin. A good guide to the flesh inside is that the paler the colour the drier and more floury the flesh, so that the darker colours will tend to be more sweet and moist.

CHOOSING AND STORING

Feel the tubers for firmness. Notice that the colour looks fresh, and discard any with soft patches. Store them in a cool dark place for a week or a little more.

COOKING

Sweet potatoes can be used for any dish or method of cooking that white potatoes are used for. But remember they are sweet, so they can be used for tarts and desserts as well. If you have not tried sweet potatoes, start with simple recipes like roasting and deep-frying and see how you like the result.

Baking for example. All the potatoes need is to be washed, pricked, then placed on the top rack of an oven at 200°C/400°F/gas6 for an hour. The crisp skins are delicious.

Or slice them thinly in a food processor or with a mandoline and deep-fry to make crisps. They make an excellent appetizer to serve with drinks.

RIGHT clockwise from left: Baked sweet potatoes, mashed sweet potatoes as a pie topping and spicy sweet potatoes (see overleaf)

Or grate them, mix them with a little chick pea flour and fry like latkes. Or boil them, mash into a purée and use them as a topping for a vegetable pie. Or add an egg to the mashed potato and fry the mixture as fritters.

Julie Sahni gives a recipe for Sweet Potato Puffed Bread in her *Classic Indian Vegetarian Cooking* – a flavoured poori, in fact, which sounds delicious. (My pooris never puff up so I have not tried it.) But see the African recipe below (Sweet Potato Puffs) for something similar.

In India the sweet potato is often used in vegetable curries. These are the recipes I find most satisfying, where the sweetness of the vegetable combines beautifully with the pungent spices. Here are two examples inspired by this tradition.

SWEET POTATO PUFFS

2 or 3 sweet potatoes
115 g / 4 oz plain flour
1 tsp baking powder
1 tsp sea salt
½ tsp ground cinnamon
½ tsp freshly grated nutmeg
2 eggs, beaten
sunflower oil for deep-frying

Boil the potatoes for 20 minutes, peel and mash, then chill for an hour.

Mix together the flour, baking powder, salt and spices. Mix the eggs with the potatoes and combine the flour and spices with the potatoes and eggs. Beat until a stiff dough forms.

Knead the dough, then roll it out on a floured surface. Cut into circles 5 mm / ¼ in thick and 8 cm / 3 in across and fry in deep oil. Use a wok as it takes less oil if you haven't a deep-fryer.

When they are puffed up and brown, let them drain. Serve them as a teacake, sprinkled with sugar, or instead of a roll for lunch.

SPICY SWEET POTATOES

2 or 3 sweet potatoes (about 900 g / 2 lb)
2 tbsp groundnut or mustard oil
1 tbsp mustard seeds
2 fresh green chillies, chopped
1 tbsp garam masala
1 tsp sea salt
2 tbsp chopped coriander leaves

Boil the potatoes whole for about 15 minutes in salted water. Drain and let them cool. Peel them (the skins will slip off easily) and slice them thickly across, about 5 mm / ¼ in thick.

Heat the oil and throw in the mustard seeds. Let them heat until they are hopping in the pan, then add the chillies and pieces of potato. Fry until the underside gets a little crisp.

Add the garam masala and salt, turning the potatoes over so that they are covered in the spices and cooking until they become crisp. Sprinkle with some green coriander and serve.

SPICY BREADCRUMBED POTATOES

2 tsp chilli powder
1 tsp turmeric
1 tsp garam masala
1 tsp sea salt
2 eggs, beaten
4 tbsp toasted breadcrumbs
2 sweet potatoes, boiled and peeled (as above)
groundnut or mustard oil for frying

Mix the spices and salt with the beaten egg. Slice the sweet potatoes thickly (as in the previous recipe), place them in a bowl and pour the egg mixture over them. Leave for an hour.

Heat the oil in a frying pan, dip each piece of potato into the breadcrumbs to cover both sides, and fry until crisp and brown.

THE ARUM FAMILY

Araceae

Nearly all this large family of climbing shrubs, herbs and marsh plants lives in the tropics. Probably stemming from the palm family, it includes ornamental plants – philodendron, monstera, arum lily and that spotted-leaf houseplant, dieffenbachia. The edible member of the family is taro; besides its tuberous root, the stems, shoots and leaves became a staple food in the Pacific very early on in history.

TARO

Colocasia esculenta

These edible tubers have been eaten for certainly all of nine thousand years – for so long that, like the onion, the original wild plant has entirely disappeared. But the colocasia genus is both huge and diverse, with varieties adapted to many different growing conditions. Common names, too, have proliferated for the same or similar vegetables: *eddo* and *dasheen* are Caribbean names. All of the plant is eaten – the immature coiled leaves and the young shoots. It is thought to have originated in India from where it travelled west, east and south; it became the staple travelling food – taro tubers were sliced, dried and smoked. Taro was also boiled and ground into a starchy paste called *poi* which is fermented. Most westerners find *poi* disagreeable, likening it to sour wallpaper glue.

Today, the humid tropical lowlands of America and south-east Asia have an agriculture and diet based on the starchy tubers, roots and rhizomes (taro, sweet potato, yam and arrowroot) which were first introduced from Asia. Paradoxically, a rice culture has intruded and taken over from the original yam and taro cultivation in those countries of South-east Asia which first regarded taro and its look-alikes (for they are all of different families) as staple food.

It is puzzling that taro achieved this status, for toxic crystals of calcium oxalate lie just beneath its skin. These produce an allergic reaction, so peel in gloves or under running water. The toxin is rendered harmless by cooking.

Taro's early success as the staple food of one third of the world has to be explained by its ease of cultivation and its huge harvest. An experiment in 1844 in South Carolina proved that one acre of rich damp soil produced one thousand bushels of taro by the second year. Some enthusiasts have spoken of its distinct artichoke-heart and chestnut flavour, but I have to confess that the first taro I ate on the island of Katatonga in the South Pacific was extremely disappointing. This had been cooked in the traditional manner in a pit called an '*umu*', dug two feet deep into the earth or sand. This hole is filled with twigs and logs and a fire is lit which is then covered with stones. When the stones are hot the taro roots, wrapped in banana leaves, are laid in the pit on top of the meat and poultry and covered in more leaves and palm fronds. The *umu* is left for twelve hours. I found the food overcooked and the taro very doughy and insipid. The one success of this feast – called '*umukai*' – was the taro leaves: they had been cooked to a purée and then mixed with minced onion and coconut milk. The islanders likened it to spinach, but it was less aggressive and tasted more like Swiss chard stalk.

NUTRITION

Rich in thiamin (vitamin B1) and other B vitamins as well as vitamin C, potassium and iron. Low in protein but a rich source of starch, the starch grains are easily digested, making them suitable for children and invalids.

VARIETIES

Taro varieties come in all sizes from large to small. The smaller ones are often called eddoes or dasheen. They are all barrel-shaped, dark brown and rather shaggy on the outside. There are also very small, smooth taros shaped like a kidney and with a pinkish bud, which is its attachment to the larger, hairier taro. These are always called eddo, though they are part of the same tuber. These are much favoured by the Chinese and Japanese in their cooking.

TARO *Colocasia esculenta*

STORING AND PREPARING

Store them like potatoes in the dark and the cool, but it is not advisable to keep them for more than a couple of weeks. They should be firm to the touch. Once peeled, keep them in salted water to prevent discolouring unless you are going to cook with them at once. Use gloves when peeling to protect against any possible skin reactions.

COOKING

I prefer to boil them in their skins; once tender, the skin peels off easily. However, cooking with the skin on sometimes causes discolouring – some varieties turn grey or a dingy lilac, changing from the original raw creaminess. This does not affect flavour and hardly matters when the taro is being used in combination with other vegetables or coated with a sauce. Taro absorbs liquid as readily as pulses, so use this quality to advantage by flavouring the liquid you are cooking the taro in: a good stock, for example, or garlic and tomato, milk and combinations of spices. Taro must also be served hot; if allowed to cool, it becomes sticky or dry and unpalatable.

Taro can be steamed or boiled, it can be chopped and added to stews where it will absorb a rich cacophony of flavours, or it can be added to vegetable soups to contribute thickness and bulk.

If steamed or boiled, taro is quite sweet and nutty, a kind of mixture of potato and water chestnut. If small and unpeeled the taro will need about 25 minutes steaming, then the skins will slip off. Boiled, the small taro need 12–15 minutes. Larger taro will need twice that amount of time, either steamed or boiled. Since it soaks up an enormous amount of fat or juice, roasting taro can be a nightmare, as it has to be continually basted as it cooks. However, deep-fried taro works wonderfully, as it crisps quickly and keeps its interior flavour and texture.

BELOW *Callaloo soup*

STIR-FRIED TARO WITH PRAWNS

2 tbsp sesame oil
2 tbsp mustard oil
2 garlic cloves, sliced
1 red chilli, chopped
55 g / 2 oz root ginger, peeled and grated
3 small taros, peeled and thinly sliced
4 large green raw prawns, peeled
1 small glass Shaoxing wine or dry sherry
1 tsp each caster sugar and sea salt
few thin spring onions, trimmed

Pour the oils into a wok and add the garlic, chilli and ginger, stir-fry for a moment, then add the taro. Fry for a minute or so until the oil is almost absorbed.

Throw in the prawns and stir-fry for another minute. Add the wine, sugar and salt and stir-fry for another minute. Serve garnished with the onions.

TARO VEGETABLE STEW

3 tbsp olive oil
3 garlic cloves, sliced
1 large onion, sliced
½ tsp asafoetida
1 tbsp oregano
2 red chillies, whole
1 large taro, peeled and diced
400 g / 14 oz tinned chopped tomatoes
1 vegetable stock cube, crumbled
115 g / 4 oz small carrots, trimmed
115 g / 4 oz small leeks, trimmed
115 g / 4 oz garden peas (frozen will do)
sea salt and freshly ground black pepper

Heat the oil in a large saucepan, throw in the garlic and onion, the asafoetida, oregano and chillies – left whole. Sauté for a moment before adding the taro, tomatoes and their juice, 575 ml / 1 pt of water and the stock cube. Bring to the boil and simmer for 15 minutes. Check that there is still some liquid; if low, add another 150 ml / ¼ pt water. Add the carrots and leeks and simmer for another 10 minutes, then add the peas. Simmer for another 3 minutes. Season.

TARO AND MUSHROOM FRITTERS

450 g / 1 lb taro, peeled and diced
225 g / 8 oz mushrooms, cleaned
1 garlic clove, crushed
1 tbsp soy sauce
1 egg, beaten
vegetable oil for frying

Boil the diced taro in plenty of water for 15 minutes, then add the mushrooms and let them steam on the top for another 15 minutes. Let the vegetables cool and drain.

Liquidize the mushrooms and mash the taro with the crushed garlic and soy sauce. Mix in the mushrooms and beaten egg.

Fry spoonfuls of the mixture in a little oil until both sides are brown and crisp.

ELEPHANT EAR

These are the shoots which are forced from taro corms; they are, in fact, the tightly coiled leaves of the mature plant. They are a great delicacy in Asia, much as we esteem sea kale or asparagus. They can be cooked in the same way, briefly poached in salted water, then drained and eaten with melted butter or hollandaise.

Alternatively they could be lightly cooked in a spicy curried sauce. Jane Grigson gives a Nepalese recipe in her *Exotic Fruits and Vegetables*.

CALLALOO

These are the leaves of the taro plant, which are poisonous if eaten raw. In the Pacific they are thoroughly boiled and then often used to wrap vegetables or meat and cooked again in an *umu* – pit cooking. Callaloo has also given its name to a Caribbean soup of salt pork, bacon, crab, prawns or shrimp with garlic, onions, okra, flavoured with chilli, lime and coconut milk. Here the leaves are shredded and cooked with the meat and not until that is cooked until tender is the fish added.

THE YAM FAMILY

Dioscoreaceae

The yam family of vines and shrubs is indigenous to tropical and warm temperate regions. They all tend to have large roots or a network of tubers like those of a potato. Very early in history, many of them were grown for their edibility, including the Chinese yam or cinnamon vine (*Dioscorea batatas*). Others have been cultivated as ornamentals for the garden – for example, black bryony, which has yellow flowers and poisonous red berries. A new function has lately appeared for the yam family: dioscorea provides a principal raw material which is used in the manufacture of birth-control pills.

YAMS

Dioscorea spp.

Each major culture throughout the world survives on a staple food. The yam, a root that can grow to a huge size, was the staple food of Africa and must have been one of the earliest foods to sustain the herbivore diet of primate and hominid, before that of humankind itself. As a whole we were unimaginative about the food and preparation of our diet in those six million years before the advent of fire. But sticks and flints abound to scrape, grate and slice roots and techniques like fermentation – burying foods so that they would half-rot, so as to become more digestible – have, I am certain, a timeless pedigree.

There are 600 species of yam; they come in all shapes, sizes and colours. Every country, almost every district and certainly every island has its own particular favourite yam which it cultivates, harvests and cooks. Yams in some Pacific islands are venerated as nature's larder, for they can be left in the ground to grow to an enormous size. On the Pacific island of Ponape the yams are described as 'four-man' and

RIGHT *Yams, taro, elephant ear and callaloo leaves*

'eight-man' – or whatever the manpower needed to lift the tuber from the ground.

All of the yams were Old World plants, except one indigenous to America, the cush-cush or Indian yam (*Dioscorea trifida*). There is also the white yam (*D. alata*) cultivated in Africa and Asia, and the Malacca yam (*D. atropurpurea*) which is eaten in Thailand. A source of confusion is that in America sweet potatoes are called yams. They are nothing of the kind, but when the Blacks were enslaved and taken to labour on the sugar plantations they mistook the sweet potatoes that grew in that part of the world for the yams they had known in Africa. The two look and taste very similar, though yams do not have the sugar content of sweet potatoes.

Yams usually look like shaggy brownish black tubers, the largest we usually encounter are the size and shape of a small marrow, but the fingered yam (*Dioscorea olata*) looks a little like a brown mitten and the sweet yam (*D. esculenta*) looks like a large baking potato covered with a few roots. In the East you might find the Chinese yam (*Dioscorea batatas*), shaped like a club, and with flowers that smell of cinnamon.

CHOOSING AND STORING

Yams must be hard, without cracks or any soft parts. They often seem of formidable size, 60 cm / 2 ft long and with a circumference of 15 cm / 6 in. They store well, so it is difficult to tell how long they have been on sale. But if you buy your yams where there is a constant demand for them, what is on sale is bound to be in prime condition. Sometimes, the yams will be cut open to show how moist and creamy the interior is. If not, scrape a yam with your fingernail to see how juicy it is. You can buy yams by the piece, according to the weight you need. Yams can be stored back home at room temperature for a few weeks.

NUTRITION

Yams are nearly all starch, but they are rich in potassium, folic acid and zinc, and vitamins B1 and 2. Potatoes have twice the vitamin C content of yams.

PREPARING

Wash the yams under a running tap then peel thickly to remove both the skin and the layer beneath. All yams contain dioscorine, which is poisonous (though cooking completely destroys it); this lies near the peel. Cut the yam into whatever size pieces you want. If not cooking at once, keep in salted water, for yam discolours easily.

COOKING

In the countries where yams are a staple food, they are generally boiled and mashed, or peeled and roasted, or left unpeeled and roasted over an open fire or barbecue. They form the starch of the meal and are nearly always eaten with fiery sauces and spicy foods.

Yam fries well, either in chips or discs. It can also be grated and mixed with flour, yeast and salt for breadmaking.

Plainly boiled yam should be served with plenty of butter or chilli sauce. Pounded yam is boiled yam beaten with a wooden spoon to let the air in and then moulded into a dough, with added salt to taste. It is served, rather like rolls, with spicy stews and soups. A tube-shaped yam of about 7.5 cm / 3 in circumference can be peeled, then cut into discs about 5 mm / ¼ in thick, parboiled for 3 minutes, then fried in olive oil and garlic. These can then be used like artichoke bottoms to pile other ingredients, well-spiced, on top, as in the recipes for kohlrabi (see page 104). Basically, use yams in any of the recipes as a substitute for potatoes, sweet potatoes or taro.

SPICY YAM SOUP

2 tbsp mustard oil
2 tbsp peanut oil
2 green chillies, chopped
2 red chillies, chopped
6 garlic cloves, chopped
2 large onions, chopped
55 g / 2 oz root ginger, peeled and chopped
900 g / 2 lb yam, peeled and chopped
water
2 vegetable stock cubes, crumbled
150 ml / ¼ pt sour cream or buttermilk
sea salt and freshly ground black pepper

Heat the oils in a large saucepan and throw in all the flavouring ingredients down to (and including) the ginger. Sauté for several minutes, stirring all the time. Throw in the yam, cook for another minute or two, then add 1.75 litres / 3 pt of water and the vegetable stock cubes. Bring to the boil and cook for 30 minutes. Season and stir in the sour cream or buttermilk.

YAM AND PEPPER SALAD

900 g / 2 lb yam, peeled and diced into 2.5 cm / 1 in chunks
1 each green, red and orange peppers, cored and deseeded
2 purple onions, sliced
12 – 15 stoned black olives
for the vinaigrette:
1 tbsp red wine vinegar
1 green chilli, chopped finely
1 tbsp Dijon mustard
5 tbsp extra virgin olive oil
1 tsp sea salt
basil, chives, parsley or dill, to garnish

Boil the yam pieces in salted water for about 20 minutes, then drain well.

Slice the coloured peppers and place in a large bowl with the onions, then add the yams and olives. Mix thoroughly.

Make the vinaigrette and pour over the salad. Leave to marinate for an hour or so.

Toss again and pile the salad on a serving dish. Garnish with any chopped fresh herbs in season.

FIERY YAM FRITTERS

450 g / 1 lb yam, peeled and chopped
2 tbsp mustard oil
5 chillies, chopped
5 garlic cloves, crushed
1 tbsp mustard seeds
1 tbsp cumin
½ tsp asafoetida
sea salt
1 egg, beaten
oil for frying

Boil the yam pieces for 20 – 30 minutes, then drain and mash.

Heat the mustard oil in a pan and throw in the chillies, garlic, mustard seeds and cumin. Cook for a moment until the chillies and garlic have browned, then sprinkle on the asafoetida.

Add to the mashed yam, sprinkle with salt and mix thoroughly with the egg.

Mould into small cakes and fry in hot oil until browned on each side. Excellent with a green salad.

THE LAUREL FAMILY

Lauraceae

The laurel family of flowering plants is characterized by its aromatic leaves and woody stems. Many reach great height as trees and are used for lumber, medicinal extracts and some of the essential oils used in perfumery. Camphor, cinnamon and cassia all come from different members of this family, which includes the bay tree (*Laurus nobilis*), a native of the Mediterranean region. The avocado is perhaps the major food plant in this family.

AVOCADO

Persea americana

Botanically the avocado counts as a fruit, so should not appear in this book at all. However, we eat it mostly at the beginning of the meal as a savoury course. It also contains more protein than any other fruit, and up to 25 per cent fat, so really qualifies for inclusion among the vegetables.

Avocados are native to the New World, found first in Mexico. The name 'avocado' comes from the Aztec word for it – *ahuacatl* – which is a shortened version of 'testicle tree'. Avocados have had a reputation for being aphrodisiac, on and off, before and since they were discovered by Columbus. A mature avocado tree, seen in the flesh, as it were, is an impressive

AVOCADO *Perea americana*

sight, often 20 metres or more / 60–80 ft high, with spreadeagled branches like a fine oak tree, but with those glossy oval leaves and pendulous fruit. A splendid example I saw in Israel reminded me more of an illuminated manuscript, a depiction of the Tree of Life; I almost expected a serpent coiled around the trunk and a nude Cranach-shaped Eve poised to offer me a pear.

It was Oviedo, historian to Charles V, who in his report to the Emperor in 1526 first described the flesh as being similar to butter. A few hundred years later one of its names was 'midshipman's butter', for avocados were taken on board ships as part of the provisions and given to junior officers. But in sixteenth-century Europe the avocado did not find favour as sweet peppers had. Though the English did plant a tree in the Botanical Gardens, Bangalore in 1819, most of Europe ignored the avocado until the middle of this century. The problem was partly one of propagation. You do not necessarily get good fruit from a plant which has grown from seed. Avocado trees must be propagated by budding and grafting from reliable parent trees, a fact which was only realized at the beginning of this century. This knowledge allowed the production of superior stock and the establishment of orchards producing fruit of uniform appearance, size and quality, a necessity when commercial production is involved.

Much work had been done in selective breeding by American botanists at the end of the nineteenth century. They worked with the three distinct types, the Mexican, with a thick skin, small fruits, anise-scented leaves and a high oil content, the Guatemalan, which has medium to large fruits with a thickish skin and a lower oil content and the West Indian, which will grow well only in the tropics but whose fruits will reach the size of a melon. These three spawned nearly 500 varieties.

In the market now you will find four varieties that have sold commercially – Ettinger, Fuerte, Nabal and Hass. The last is the small knobbly black avocado that gave the common name 'alligator pear' to the fruit. Ettinger and Fuerte are pear-shaped and ripen early in the year. Nabal is rounder and is picked from the middle of January to June. The avocado is notable for staying on the tree fully grown, but not finally ripening for nine to ten months. The Hass can stay on the tree from October until the following June.

Avocados are now grown all over the world in tropical and sub-tropical countries, though they are not partial to wind or too dry a climate. They thrive in Central and South America, Florida and California, Hawaii, the South Pacific, Australia, South Africa and the Canary Islands, as well as Israel and other Mediterranean countries.

Today Israel exports something near to 100,000 tons of avocados. The first seeds were planted in Palestine in 1895, but it was not until 1956 that Israel began to farm the fruit commercially. Growing avocados requires huge initial outlay as there is no fruit for the first six or seven years, but then trees can go on bearing for up to 80 years. In Mexico, it is not uncommon to have trees which are over 150 years old.

The tree produces its fruit once every two years. It flowers every other year as the fruit stays on the tree for so long. The avocado begins to ripen 24 hours after it is picked. It will then take four to ten days at a temperature of around 5°C/42°F to reach perfection.

The avocado butter tradition continues in all the countries where avocados are cheap and plentiful. Spread on bread or tacos, it makes a cheap and nutritious snack, especially popular with children.

Jane Grigson tells us that the avocado had crept into very grand English grocers in the 1930s before the Second World War. But I remember it first in the 1950s. I had my first avocado in about 1955, at a restaurant outside Brighton, who served it peeled and fanned out on the plate with a vinaigrette sauce. This method of serving was entirely unknown and did not become trendy for another thirty years. I ate my first avocado soon after James Bond ate his, for in Ian Fleming's first Bond book, *Casino Royale* (1953), Bond chooses to conclude the meal with half an avocado pear, while his girlfriend ate the strawberries.

NUTRITION

Avocado contains some dietary fibre and a large amount of potassium, some carotene and vitamins E, C and B6 as well as thiamin and riboflavin. Its protein content is as high as its carbohydrate content. The fat it contains is mostly monounsaturated. The calorie content is lower than many people think – it equals that of one roast pigeon.

CHOOSING AND STORING

Avocados are displayed either as unripe, ie hard, or as ripe, when they are soft and sold cheaper. The unripe ones will ripen within 4–10 days at room temperature; once that has happened the avocado will keep in the refrigerator for a few days more.

An old Israeli farmer I spoke to wanted all British families to have one or two avocados ripening in their kitchens all the time, as they did. I pointed out to him that they were rather more expensive for us and it was unlikely to happen.

If you choose to buy a ripe avocado, first inspect it for any patches of brown or black coloration and refuse to choose one which is not green and gives slightly to gentle pressure. The Hass avocado (the black knobbly one) does not show any discoloration, so how soft it feels is the only sign of ripeness. But serious errors can be made with Hass, as the skin is so thick it does not always give enough when gently squeezed. I have returned home to find that on slicing the fruit in half, some of it is black and inedible.

PREPARING

Cut a ripe avocado in half, dig out the stone with a teaspoon and fill the central cavity with a garlicky vinaigrette. That is the simplest, and a quite delicious, method of enjoying this fruit. But the cavity can also be used to hold shrimps, prawns, crab or even taramasalata. It can also hold various salad ingredients in a tomato and chilli sauce – like chopped cucumber and spring onion. One can be highly imaginative and also vary the type of sauce: diced potato and sour cream is heavenly. In fact, ripe avocado flesh

BELOW *Various types of avocado*

can be so enhanced that it becomes one of those ambrosial delights when any ingredient or flavouring that is a little sharp and acidic is added.

Avocados can also be peeled and sliced either across or lengthwise. Once this is done they must be sprinkled with lemon juice, for the flesh oxidizes quickly and starts to go black. Grind a little coarse sea salt over just before serving. It is the above method of serving avocado at a party as an appetizer that I came across in Israel. It is simple but good.

Avocado flesh can also be diced and added to a vinaigrette for a green salad – the avocado pieces will coat the leaves. If you add chopped hard-boiled egg, onions, capers and garlic croutons, this becomes a delicious summer salad meal.

Diced avocado added to an onion and potato salad is also excellent, as are young spinach leaves and diced avocado, as are apple, avocado, walnuts and watercress. The salad choices are almost infinite; because avocado has 25 per cent protein, these salads are sustaining and nutritious.

Purées can easily be made from the flesh once it is ripe. In this state it is a staple ingredient in a host of different sandwiches. Particularly good mixtures include avocado and rocket; avocado, onion and pickled gherkin; avocado and that paprika cheese Liptauer; or avocado and any hot Indian chutney.

In Israel they scoop the flesh out of several ripe avocados, add salt, pepper and lemon juice, mash thoroughly together and keep in an airtight jar in the refrigerator for use as a spread. As long as there is no air in the jar, it will keep for a few days without discolouring. Avocado purée can also be mixed in with mayonnaise and used as a sauce for crudités.

Soups are simplicity to make, as are sauces. The soups should be iced and the sauces go well with fish or vegetable dishes.

COOKING

Strictly this is inappropriate, for avocados must never be cooked. But they can be served hot. The timing and the temperature are important: if the avocado begins to cook, it becomes bitter and unpleasant.

However, avocado halves can be filled with a hot sauce and placed in a hot oven for about three minutes, then served. This just gives the avocado enough time to become hot, but *not* to start cooking.

Alternatively, lay slices of avocado on hot toast, then cover with strips of mozzarella and place under a hot grill so that the cheese melts. This is particularly good as the avocado is barely warm, but one achieves a pleasing complexity of texture as well as temperature. For sparkle, first spread a tiny smear of harissa or a hot chilli sauce on the toast.

ICED AVOCADO SOUP

for 6

1 large cucumber, chopped
3 green peppers, cored and deseeded
1 garlic clove, crushed
generous handful of fresh parsley, chopped
3 ripe avocados, stoned and peeled
1.1 litres / 2 pt soya milk
300 ml / ½ pt buttermilk or sour cream
sea salt and white pepper

Blend the cucumber, peppers, garlic and parsley to a thick purée in a food mixer. Add the avocado and blend again, season and add the soya milk gradually.

Finally stir in the buttermilk and sour cream. The soup should have the consistency of double cream.

Place in the refrigerator for a couple of hours before serving.

CHILLI AVOCADO CREAM

2 green chillies, chopped, seeds and pith removed for a milder taste
3 tbsp olive oil
3 ripe avocados, stoned and peeled
150 ml / ¼ pt double cream
sea salt

Heat the olive oil in a pan and fry the green chillies for a moment until they just brown a little, then place in a blender jar and liquidize. Add the avocado flesh, salt and double cream. Blend to a thick purée. Use as a sauce for large prawns or poached fish, or as a dip for crudités.

GUACAMOLE

This traditional Mexican recipes to be eaten with taco chips achieves the best results if you refuse to use the blender jar or food mixer and mix by hand instead.

3 ripe avocadoes, stoned and peeled
1 or 2 green chillies, finely diced
3 or 4 ripe tomatoes, skinned and finely diced
1 lime
1 garlic clove, crushed
150 ml / ¼ pt olive oil
2 tbsp chopped fresh coriander

In a large bowl, mash the avocado flesh coarsely and add to it the chillies, tomatoes, lime zest and juice and the garlic. Mix thoroughly.

Add the olive oil slowly, beating it in to the mixture. Finally, stir in the chopped green coriander.

STIR-FRIED AVOCADO SALAD

2 tbsp sesame oil
2 tbsp mustard oil
2 garlic cloves, sliced
2 green chillies, finely diced
115 g / 4 oz baby carrots, trimmed but left whole
115 g / 4 oz baby leeks, trimmed but left whole
115 g / 4 oz calabrese florets, trimmed and separated
75 ml / 3 fl oz dry sherry or vermouth
1 tbsp soy sauce
1 tsp caster sugar
2 ripe avocados, stoned and peeled
5 leaves Webb's lettuce
2 tbsp chopped fresh coriander

In a wok, heat the two oils and throw in the garlic and the chilli. Fry for a moment, before adding the vegetables. Stir-fry the carrots, leeks and calabrese for about a minute, then add the sherry or vermouth and the soy sauce and sugar. Stir-fry for another minute, then turn the heat off.

Slice the peeled avocado lengthwise and carefully mix in with the vegetables, so that the pieces are covered with the sauce but do not break up.

Arrange the leaves of lettuce over a platter and gently tip the wok so that the contents pile up in the centre. Sprinkle with the coriander. Serve at once, while still warm, or eat when cooled. An excellent summer luncheon dish.

HOT STUFFED AVOCADO

225 g / 8 oz potatoes, peeled, boiled and diced
2 tbsp olive oil
1 green chilli, diced
1 garlic clove, chopped
1 tin of anchovies
150 ml / ¼ pt sour cream
55 g / 2 oz Parmesan cheese, grated
2 ripe avocados, stoned
freshly ground black pepper

Preheat the oven to 190°C/375°F/gas5.

Heat the olive oil in a pan and add the chilli and garlic. Fry for a moment or two, then add the diced potato and the anchovies in their oil. Move the potato around so that it is covered with the spices and the anchovy. Let it brown a little and add the seasoning. Then stir in the sour cream. Remove from the heat.

Place the two halved avocados on a baking tray and fill the cavities with the stuffing, making sure you cover all the exposed avocado. Sprinkle with the Parmesan.

Slip the tray in the oven for 3 minutes and serve at once.

Fungi

MUSHROOMS

AND

TRUFFLES

Mushrooms

CULTIVATED MUSHROOM *Agaricus bisporus*
FIELD MUSHROOM *Agaricus campestris*
PARASOL MUSHROOM *Macrolepiota procera*
OYSTER MUSHROOM *Pleurotus ostreatus*
CEP *Boletus edulis*
CHANTERELLE *Cantharellus cibarius*
GIANT PUFFBALL *Langermannia gigantea*
MOREL *Morchella vulgaris*

Truffles

BLACK or PÉRIGORD TRUFFLE
Tuber melanosporum
SUMMER TRUFFLE *Tuber aestivum*
WHITE or PIEDMONT TRUFFLE *Tuber magnatum*

Mushrooms and their allies belong to a fundamentally different group from the other vegetables in this book, all of which are produced by chlorophyll-based 'higher plants' which flower and fruit in their myriad different ways. Modern biologists classify the fungi – which lack chlorophyll and reproduce by means of spores, not seeds – in a separate 'kingdom' of their own, on a par with the plant and animal kingdoms.

MUSHROOMS

It is not surprising that a certain magical aura surrounds mushrooms and other fungi. Not only are some hallucinogenic and others poisonous, but they all appear suddenly as if from nowhere. The first writer to mention fungi is Theophrastus (*c.*300 BC), who describes truffles as having 'neither root, stem, branch, bud, leaf, flower, nor fruit'. If that is not enough, he goes on to say that nor do they have 'bark, pith, fibres, nor veins'. Pliny sounds more enthusiastic: 'Among the most wonderful of all things is the fact that anything can spring up and live without a root.' But suspicions were never quite entirely erased. Mushrooms were thought to be engendered by thunder and to grow where lightning struck the ground. Other explanations for the appearance of mushrooms included the supposed venom of toads, and some 'evil ferment of the earth'.

However, there is little doubt that various mushrooms were much enjoyed as food in Ancient Rome. Special silver vessels (*boletaria*) were designed for cooking them, and their delights were celebrated in verses and epigrams. In the last half of the second century BC, when the Sumptuary Laws came in to curb the excessive use of meat and fish at banquets, mushrooms and other edible fungi became a favourite gastronomic delicacy.

The ideas about the generation of mushrooms formed in the classical world remained throughout the Middle Ages, until in 1679 Malpighi decided that fungi have their own 'seed' and that also 'they can sprout from the growth of fragments of themselves'. A little later, in 1707, J. P. de Tournefort discovered mycelium. 'According to appearance these white threads are none other than the developed seeds, or germs of mushrooms.' It was Tournefort who began the cultivation of mushrooms of the *Agaricus* genus, which includes field and horse mushrooms and their close relatives. His paper given to the French Academy described the method. Ridge beds were prepared in the open from turned stable manure encased with soil and inoculated with pieces of mouldy horse manure. Tournefort believed that stable manure always carried invisible mushroom 'seeds', which in a few days would reveal themselves as delicate white filaments and have '*une odeur admirable de champignon*'.

The idea crossed the Channel to London, where Philip Miller reproduces the method in his *Gardener's Dictionary* (1731). He advises that if there is a dearth of mushroom spawn, you should go out in August and September and then open the ground around mushrooms to find 'white knobs which are the off-sets or young mushrooms'. Stable yards would generally have a shady corner given over to the hoped-for growth of mushrooms, and the land-owning aristocracy had mushroom beds built in darkened outhouses. The whole process was an uncertain business, however, and it took until the end of the nineteenth century, when 'pure-culture spawn' was developed in 1894, for mushroom growing to begin to develop into a commercial industry.

In France, the disused underground limestone quarries in the Paris region became the centre of mushroom cultivation in the mid-nineteenth century – hence the name '*champignons de Paris*'. With the patenting of the method using pure-culture spawn, France had a monopoly which enabled them to become the leading mushroom producers for almost twenty years.

It seems somewhat astonishing that the West had not learned from the Far East long before this, for they had been cultivating *shiitake* (*Lentinus edodes*), which grows on the bark of logs, for two thousand years both in Japan and China. In the second century BC mushroom cultivation is described in *The Pharmacopoeia of the Heavenly Husbandman*. The padistraw mushroom (*Volvariella volvacea*) is also widely cultivated in South-east Asia.

Evidence of the use of mushrooms as food has been noted all around the globe, from the *Polyporus mylittae* eaten by Australian aborigines, the large *Poria cocos* made into bread by North American Indians and the 'globular, bright yellow fungus' (*Cyttaria darwinii*) noted by Darwin in 1834 growing at Tierra del Fuego, which was eaten by the natives. These and scores of other edible fungi are not cultivated, but gathered in season in the wild. For all the sophistication of contemporary science, which has helped us

---

MUSHROOMS AND TRUFFLES *Fungi*

understand the complex processes by which fungi grow, and the conditions and specific hosts they require, most of the prime edible species continue to resist various attempts to cultivate them. At a time when produce from all around the world appears constantly and consistently on supermarket shelves, wild mushrooms retain a degree of their ancient mystique. This is due partly to the intrinsic unpredictability of their fruiting, which depends very much on suitable weather conditions, but also not least to the fact that collectors like to preserve an aura of secrecy around their hunting grounds.

There are many thousands of different fungi – new ones are continually being discovered; and they have always had an equivocal relationship with humankind. They play a vital role as agents of decay, as one of the elements that help to maintain the earth's biosphere; they have been a source of drugs and are used commercially today in medicines. They can attack plants and decimate harvests (a fungus is responsible for the dreaded potato blight), but they also play a part in ensuring the maintenance of soil fertility by breaking down organic residues.

Among those important in foodstuffs – which include the yeasts responsible for fermentation and the moulds that produce cheeses – we vegetable consumers are most concerned with the 'higher fungi'. These are the species in which the fruiting body of the mushroom – the reproductive part of the organism, which distributes the spores (carried under the cap on the gills of the familiar 'mushroom-shaped' kinds) – is large and succulent enough to be tasty, but at the same time lacking any substances that are harmful to humans.

Only a few mushrooms are poisonous; many more are tasteless or impossibly tough, and hardly worth gathering for cooking, but a few dozen are of such enormous gastronomic worth that an autumn mushroom hunt in your local woods and pastures is almost a necessity.

A handful of species can be cultivated and are routinely available in shops. They all have their counterparts and close relatives in the wild. The shop-bought ones have the virtues of availability and consistency, but their flavour sometimes verges on blandness. A reliable tactic is to use the cultivated mushrooms in bulk for texture, and to augment them with a few precious wild mushrooms – even reconstituted dried ones – for added flavour.

NUTRITION

Mushrooms contain vegetable protein, vitamins B1 to 6, phosphorus, potassium, sulphur and folic acid.

CHOOSING

Whether your source is a supermarket, a market stall, or the fields and woodlands, select young, firm, fresh specimens without bruising or discoloration.

It is worth investing in a couple of good field guides to help identify mushrooms picked in the wild, but never eat fungi you have gathered yourself without having their identification confirmed by an expert. In some countries pharmacists or public health officials are authorized to do this and will have checked mushrooms on sale on market stalls.

Avoid gathering mushrooms at busy roadsides or on old mining or industrial sites, because mushrooms tend to concentrate any toxic substances such as heavy metals that occur in their vicinity.

STORING

Don't allow mushrooms to steam and sweat in a polythene bag on the way home, whether they are wild or cultivated ones: they travel best in a basket, lightly wrapped or covered. Once home, most varieties will keep in the fridge for a few days at most. Mushrooms do not freeze well in their natural state. If you have a glut – especially of wild mushrooms, which are far too precious to waste – it is best to preserve them in some way such as by drying, pickling, preserving in oil or cooking and then freezing.

A good holding-pattern for a large haul of fresh mushrooms of any kind is to clean them and sauté them in butter and/or oil as soon as possible: even if not eaten straight away, they keep better for a day or so in this state and can also be frozen.

Dried mushrooms are increasingly available, at a price. Choice slices of favourite species such as ceps, in special packets, are the most expensive, but you can buy small quantities by weight in many stores.

Since their merit lies in taste rather than texture, it is not vital to have pristine slices: the cheaper oddments taste just as good. It is worthwhile drying even scraps of any ceps or morels that you find for this reason. Reconstitute dried mushrooms by pouring boiling water over them and leaving them for half an hour. Use the soaking water whenever possible (for the same recipe, or save it for stock or soup): it contains a great deal of the flavour, but do strain it carefully to eliminate any grit.

DRYING YOUR OWN MUSHROOMS

Slice the mushrooms, especially fleshy ones like closed-cap field mushrooms and ceps, into pieces no more than 1 cm / ½ in thick. One method is to place clean newspaper over the top of a solid-fuel stove and leave the mushrooms laid out on this overnight. However, any dry heat will do – in an airing cupboard, on a wire tray over a radiator, or in a *very* low oven with the door left slightly ajar. When completely dry and shrivelled, bottle the mushrooms in airtight jars and make sure the lids are screwed on tightly. Dried mushrooms will keep for years.

They can be added to soups and stews as they are, but generally it is best to reconstitute them by pouring boiling water over them and leaving them for half an hour. Use the soaking water whenever possible (for the same recipe, or save it for stock or soup): it contains a great deal of the flavour, but do strain it carefully to eliminate any grit.

PREPARING

Wipe or dust off dirt particles with a soft brush rather than washing mushrooms (you need to avoid increasing their water content). There is no need to peel most types. Cut off any stem bases that are earthy. Follow any specific instructions for individual mushroom species: some have tough stems that are best discarded, but in common mushrooms, ceps and chanterelles the stems are as delicious as the caps.

LEFT Cultivated mushrooms, chestnut, field and oyster mushrooms

Also note any special preparation needed to make a particular kind of mushroom palatable. Some wild kinds must be cooked because they are toxic raw; others are digestible only after blanching (and the water must be thrown away).

COOKING

To savour the special flavours, simply frying in butter and olive oil is unbeatable. Pepper and salt are necessary; parsley, garlic, shallots, cream and other additions optional.

Mushrooms are largely water, which makes them tend to shrink during cooking and to exude lots of liquid – sauté briskly to avoid the risk of them stewing and becoming tough; you can pour off the juices that seep out at the start of cooking, then add them back to the pan to evaporate when the mushrooms are sizzling merrily, or add them to an accompanying sauce.

Recipes involving mushrooms as fillings, sauces and so on are virtually interchangeable, which is just as well when the occurrence of the wild varieties is so unpredictable. A small number of wild mushrooms – or a few dried ones from the store cupboard – can be relied on to perk up the impact of a dish in which the bulk is supplied by blander cultivated kinds.

THE CULTIVATED MUSHROOM

Agaricus bisporus

This is the mushroom that has been cultivated in Europe and America for some 150 years. It is occasionally found growing wild and a number of close relatives – other, wild *Agaricus* species – are very similar in appearance, including the field and horse mushrooms. All are the familiar umbrella shape, white-fleshed, with pinky-beige gills when young maturing to deep brownish-black.

In the shops these come as button mushrooms, closed cap, open cap and flat mushrooms. They are merely the same mushroom at different stages of growth. The whole fruit-body is edible, apart from any earthy stem base. There is no difference in flavour between the different types. All will shrink in size when cooked as they have a high water content. They can, alternatively, be eaten raw, or left to marinate in a vinaigrette and served as a salad – in my view, one of the tastiest methods of dealing with cultivated mushrooms.

The larger flat ones can be grilled or baked, gill side upwards, with a little garlic butter or oil. They can also be stuffed. These also make excellent soups and sauces.

CHESTNUT MUSHROOM: This is a form of *Agaricus bisporus* with a thicker stem and a pleasant mahogany-coloured cap exterior, so called because it has a slightly nutty flavour. It cooks well and if sliced through in half, stalk as well, will keep its shape and look enticing in stir-fried Chinese dishes.

FIELD MUSHROOM

Agaricus campestris

These wild cousins of the cultivated mushroom have a wonderful flavour and aroma. Found in meadows and fields, they open to large, flat caps with very black gills when mature. You can often find them in the autumn in farm shops.

These also lose a considerable amount of moisture and so shrink when cooked. They are best fried in oil or butter with a little garlic. So excellent are they like this I could not endure to use them in any more complicated manner.

PARASOL MUSHROOM

Macrolepiota procera

This wild mushroom is shaped exactly like a pointed parasol or umbrella when fully grown. The cap surface is buff-coloured or grey-brown with darker shaggy scales and the gills beneath are white. The closed young parasol is completely egg-shaped. The stem grows very long and often becomes too fibrous to eat. I have seen parasols growing near me with caps a foot in diameter. Sometimes fully opened parasols have become desiccated and tough, but before they reach this stage they are very good to eat. They can be fried in oil or butter or dried and stored.

OYSTER MUSHROOM

Pleurotus ostreatus

This is one of the most delicious mushrooms, first noted and drawn in 1601 by the botanist Clusius as part of a collection of 86 watercolours of various mushrooms found in Austria and Hungary. Now cultivated, it is widely available (though on the expensive side). The mushrooms grow in clumps on rotting wood. The cap, stem and gills are all the same colour – a pale buff-grey, sometimes faintly tinged with blue or pink. A yellow species of pleurotus is also often available on sale.

I like to tear oyster mushrooms in strips (the stems are tough, particularly in larger specimens and must be thrown away) and to add them to stir-fry dishes. But if you are feeling luxurious, cook them as a dish by themselves, fried in sesame and olive oil with garlic and baby sweetcorn. With good crusty bread to mop up the juices and a green salad it makes a marvellous meal.

CEP

Boletus edulis

Known as '*porcini*' in Italian and '*cèpe*' in French, this is one of the best wild mushrooms. With a rounded cap shaped just like a penny bun atop a stout stem, it grows on the ground in woods all over Europe. It is easily identified, as all the boletes have a mass of sponge-like tubes beneath the cap rather than gills. Outside Britain, ceps can be found in the autumn in

RIGHT Various types of wild mushroom

markets. Much of the crop is dried and sold commercially. Dried porcini from Italy tend to be expensive, but you need very little to achieve wonderful results. As with oyster mushrooms, the flavour is so good it is a mistake to swamp ceps with other ingredients and flavours. They are best simply fried in butter and olive oil, with the addition of a little parsley. Jane Grigson in *The Mushroom Feast* gives a host of inviting recipes. Both cap and stem are delicious – only the earthy stem base needs to be trimmed away. But except with very young ceps where the tubes are still firm and creamy white in colour, it is often best to scrape away the spongy mass of tubes beneath the flesh of the cap: they can become unappetizingly soggy when cooked.

CHANTERELLE

Cantharellus cibarius

This is a funnel-shaped wild mushroom which is easily recognized as the gills (actually gill-like wrinkles rather than true gills) run from the underside of the cap straight into the stem, like the ribs of fan vaulting. It appears on the ground in woodland areas. Their colour is striking – ranging from a pale cream to deep egg-yolk yellow. The entire mushroom is edible – just discard the base of the stem. Use a soft brush to dislodge specks of dirt from between the gills and from folds in the cap. Their flavour is excellent, fried in butter with chopped shallots, with a little tarragon or parsley. Chanterelles can also be bought dried or dried at home and are easily reconstituted with boiling water in the usual way.

GIANT PUFFBALL

Langermannia gigantea

By far my own favourite, this creamy-white spherical fungus can grow larger than a football. Eventually it matures to a brown sponge-like ball which puffs out its billions of snuff-coloured spores. Harvest puffballs while still pure white, from the size of a large grapefruit upwards. The outside has a velvety texture, and the whole fruit-body should feel fairly firm: if it feels spongy the puffball will already have gone too far. The critical test comes when you cut it open: if it is completely creamy white inside, you can eat it, but once it turns yellow deepening to a sandy buff, you should discard it.

If you catch a puffball in its prime, what a treat you have in store. Slice it like a large loaf into pieces about 1 cm / ½ in thick and fry in olive oil, butter, a little garlic. They will shrink, but not that much; fry until just tinged with gold. I once served them like this as appetizers at a party and foxed everyone as to what they were.

A puffball will keep for several days even after half of it has been sliced away, if covered in cling film and kept in the refrigerator. Watch it carefully for signs of colouring: as soon as it begins to turn yellow, it becomes inedible.

MOREL

Morchella vulgaris

This and other morel species are among the great edible fungi – in Scandinavia they call them the 'truffles of the north'. Morels are seen in the spring – March to May – and tend to grow on loose sandy soil overlying chalk. They also like burnt areas, so they can be an unexpected fruit from forest fires.

Morels are roughly conical in shape, with a deeply crinkly sponge-like exterior. They are hollow inside, and all this convoluted shaping means that earwigs and other insects lie in their recesses. Thorough cleaning is vital. Cut the morels in two and wash quickly under a running tap, inspecting the crinkly tops and interior crevices for any persistent insect life. (Brief washing is acceptable for morels – an exception to the general rule.) Morels can be dried by threading and hanging up in a warm kitchen for a couple of days, then storing in an airtight container.

RIGHT *Wild mushroom salad (overleaf)*

To reconstitute, either soak in water for half an hour or add directly to cooking if the dish has liquor or a sauce and will continue to cook for ten minutes.

Recipes for morels are often quite rich; they are teamed with cream and egg yolk, sherry and buttery pastry. Fine, if your digestion will endure it. But I prefer all my fungi done in the simplest manner; their flavours are so striking and so satisfying, they need very little added. If you feel like experimenting, try some of the morel recipes in Jane Grigson's *Mushroom Feast*, that wonderful source book on the cooking of all mushrooms.

WILD MUSHROOM SALAD

This is an excellent method of enjoying the harvest from a mushroom hunt. Choose young ceps, chanterelles, field mushrooms and parasols. Wipe them clean, discard the stalk bases and cut in half or slice thickly if large. Blanch them in boiling water and leave for 30 minutes.

Drain, dry, then pack them into a glass jar and cover with extra-virgin oil flavoured with the grated zest from a lemon and 2 crushed garlic cloves. Leave the jar in the refrigerator for one or two days.

Fill a platter with the leaves from Webb's or cos lettuce, take out the fungi with a slotted spoon and lay them on the lettuce, sprinkle with balsamic vinegar and sea salt. (Use the olive oil that is left over for a vinaigrette.)

MUSHROOM PÂTÉ

This is best made with field mushrooms. If these are not available, use the flat open cultivated kind, with a little dried porcini or any other dried wild variety.

85 g / 3 oz butter
1 red chilli pepper, deseeded and finely chopped (optional)
1 bay leaf
450 g / 1 lb field mushrooms, chopped small
55 g / 2 oz reconstituted dried ceps (porcini), chopped small
2 tbsp soy sauce
85 g / 3 oz fresh white breadcrumbs
sea salt and freshly ground black pepper

Melt the butter in a pan, add the chilli, if using, and the bay leaf, followed by the mushrooms and porcini with their soaking water. Season and fit a lid to the pan. Leave over a gentle heat to cook slowly for 15–18 minutes.

Remove from the heat, and add the soy sauce, seasoning and breadcrumbs. Mix together thoroughly and leave to cool to room temperature.

Pour into a mould or terrine, smooth out the surface and place a weight on top. Chill in the refrigerator for 24 hours.

This gives you a fairly chunky texture. If you desire a smoother pâté, then purée the cooked mushrooms with their liquid in a blender before adding the breadcrumbs.

PORCINI SAUCE

This sauce is wonderful served on fresh pasta, polenta or rice.

55 g / 2 oz dried ceps (porcini)
55 g / 2 oz butter
115 g / 4 oz cultivated mushrooms, diced
30 g / 1 oz plain flour
150 ml / ¼ pt white wine
1 tsp soy sauce
300 ml / ½ pt vegetable stock
sea salt and freshly ground black pepper

Reconstitute the porcini by pouring enough boiling water over them to cover, and leave them to soak for 30 minutes.

Melt 30 g / 1 oz of the butter in a pan, throw in the cultivated mushrooms and stir over a medium heat for 2 minutes. Add the porcini with their soaking water and sprinkle with a pinch of sea salt. Put the lid on the pan and let them sweat for 5 minutes. Leave to cool and then liquidize.

Make a roux with the remaining 30 g / 1 oz butter and the flour, season and add the rest of the ingredients. Bring to the boil and let the sauce thicken, then add the porcini and mushroom purée. Mix them in thoroughly.

Reheat when you need the sauce.

MUSHROOM RISOTTO

You can use cultivated mushrooms for this dish, but it is best if made with a mixture of field and oyster mushrooms.

55 g / 2 oz butter
3 tbsp olive oil
2 – 3 shallots, diced
225 g / 8 oz mushrooms, chopped
140 g / 5 oz arborio rice
pinch of saffron strands
150 ml / ¼ pt white wine
300 ml / ½ pt vegetable stock
55 g / 2 oz Parmesan cheese, grated
sea salt and freshly ground black pepper

Melt half the butter with the olive oil in a saucepan, add the shallots and the mushrooms, and let them cook gently for a few minutes.

Add the rice and saffron; stir and mix thoroughly, letting the rice soak up the juices, then add the wine, seasoning and the vegetable stock. Bring to the boil, then simmer with the lid on the pan for 8 minutes.

Remove from the heat and let it stand for a further 5 minutes. Melt the remaining 30 g / 1 oz butter and stir it in with the Parmesan before serving.

FILO MUSHROOM PARCELS

Filo pastry is obliging, as long as you work quickly, and can be made into any shape you care for – a purse, where the tip is pinched together, a triangular *samosa* shape, or a roll like a *dolmades*.

2 sheets filo pastry
1 garlic clove, crushed
2 tbsp olive oil
1 egg, beaten
1 tbsp sesame seeds
for the filling:
30 g / 1 oz butter
55 g / 2 oz reconstituted dried ceps (porcini), diced
225 g / 8 oz cultivated mushrooms, diced
1 glass of red wine
pinch of ground coriander
1 bay leaf
1 bunch of spring onions, chopped
sea salt and and freshly ground black pepper
150 ml / ¼ pt sour cream

Preheat the oven to 200°C/400°F/gas6. Defrost the filo pastry.

Make the filling: melt the butter in a pan and throw in both kinds of mushrooms and the chopped onions with the wine, coriander, bay leaf and seasoning. Leave on a low heat for 10 minutes, then take the lid from the pan and raise the heat to drive off all the liquid: this should take about 2 minutes. Leave to cool and mix in the sour cream.

Unroll 2 sheets of filo pastry. Mix the garlic with the olive oil and paint the sheets with the mixture. Cut the sheets into 20 × 10 cm / 8 × 4 in strips.

Place a tablespoon of the mushroom mixture at the end of a strip, leaving 1 cm / ½ in clear at either end, then begin to roll up the pastry, tucking in the ends as you go.

Place each one on a baking tray, the end of the roll downwards. When all are done, paint each one with beaten egg and sprinkle the sesame seeds over the top.

Bake in the oven for 10 minutes or until they have puffed up a little and turned golden brown.

TRUFFLES

These are the stuff of legends. Before I had ever even heard the name, I smelt the astonishing, pervasive aroma. In the early fifties a friend drove back from France with a truffle in a matchbox which was in her handbag. Not only the matchbox, and the handbag too, but also the whole interior of the car smelled of the most enticing, appetizing fare – an intensity of pong that I breathed in as if it were life-saving oxygen. I enquired what it was and began to learn about truffles. It seemed to me a miracle and deeply maddening that this intensely aromatic fungus lay hidden, undetected by us just beneath the earth. I remember that she cooked me a simple risotto with this precious object made up of merely a little butter and chopped shallot, then some white wine, rice and slices of the truffle. It was heaven.

Like all good things, the truffle has a long and honourable history. Theophrastus describes it as a vegetable, descended from lightning and capable of reproducing itself by seed. Plutarch ponders on its source as being a chemical reaction involving soil, water, heat and lightning. Others believed that the truffle grew out of the sting from a fly on the root of a tree. The Romans thought that the best truffles came from Africa. Juvenal writes: 'Lybians, unyoke your oxen, keep your grain, but send us your truffles.' He also advises Romans to prepare them with their own hands, for they are too precious to leave to the servants.

It was John Ray in the seventeenth century who classified the truffle as a mushroom. Later, in 1851, Edmond Tulasne (described as the founder of modern mycology) uncovered the fact that the truffle's mycelium surrounds the roots of a host tree and forms what is now known as a mycorrhizal association. (This symbiotic relationship occurs with various other mushrooms including ceps and chanterelles.) It was realized that truffles always grow near oaks – deciduous oaks in some areas, evergreen and cork oaks in others (in Spain, for example, where I have collected and eaten them). Since truffle and tree form a symbiotic relationship, truffles cannot be grown independently of the host.

In recent years in France crops of oaks have been planted in soil which has been impregnated with truffle mycorrhiza, and truffles are being produced within five years.

BLACK or PÉRIGORD TRUFFLE

Tuber melanosporum

Périgord truffles are ebony-black (though the immature fungus is a deep, dark red), knobbly, and can have a diameter anything from 2.5 cm / 1 in to 15 cm / 6 in. Truffles should be succulent, but firm. At the beginning of this century when truffles were reasonable in cost, they were sometimes baked whole in the oven like a potato. But like everything else, truffles vary in quality. The worst ones are innocuous and insipid, while the very best are so astonishing in their flavour that the moment you eat them remains quite unforgettable. Waverley Root tells of his experience

BELOW *Black truffle*

when he felt that all they had was 'a faint liquorice flavour'; later he saw truffles 'as large as tangerines, almost black with a suggestion of purple, attractively pebbled and glistening as though they had just been oiled'. When he finally eats one he is rendered almost speechless: 'My mouth was flooded with what was probably the most delicious taste I have ever encountered in my entire life, simultaneously rich, subtle, and indescribable.'

So do not judge black truffles by those inedible bits in tinned foie gras: wait until you have a chance to buy fresh ones in a market or go on a truffle hunt and find them for yourself. In 1892, 2,000 tons of truffles were harvested, but the numbers had already declined by the beginning of the First World War – only 300 tons were collected from Périgord. Now, France produces only 25 to 150 tons annually. The season for harvesting them runs from November to March. The scent of the truffle contains the highest pheromone content of any plant (the reason why that aroma seduces one's sensibilities). The sexual stimulus is what excites the sow and the dog to dig for the object. Sows can only be used up to about seven months old, for after that they become far too strong for their handler, and the sow cannot be held back and will then eat the truffle. Terrier dogs have to be trained early on by smearing a truffle on their snouts and when they have dug down to the truffle – generally a foot beneath the surface – they are quickly rewarded with a favourite food and the truffle goes into the bag.

SUMMER TRUFFLES

Tuber aestivum

This is the type of truffle which grows in Britain. They should be looked for in late summer in chalk beech woods just under the surface and occasionally just showing through. The last British truffle hunter, Alfred Collins, retired in 1930. He searched in the Winterslow area of Wiltshire, where truffles had been harvested for 300 years, with the help of two trained dogs. On a good day he could collect 11 kg / 25 lb. In

ABOVE A pig snuffling for truffles

1920 he put the price up to two shillings and sixpence a pound – twelve and a half pence in decimal money. His father once found a truffle weighing 900 g / 2 lb and he dutifully sent it to Queen Victoria, who replied saying she would send him her portrait. Later, a single gold sovereign arrived with her image stamped upon it.

WHITE or PIEDMONT TRUFFLE

Tuber magnatum

This is the famous white truffle of Bologna and Northern Italy. It is larger than the black and many people prefer its flavour. I tasted it in the late fifties when I was in Florence and contacted Harold Acton. I had met him once in London and he had admired a short story of mine published in *The London Magazine*. He invited me to tea at his seventeenth-century villa – la Pietra. The tea turned out to consist of very dry dry martinis and paper-thin white truffle sandwiches: an amazing combination. Excellent Italian bread, unpasteurized butter and slivers of white truffle.

My next most memorable white truffle experience was thirty-five years later, when Anton Edelmann of the Savoy cooked me a white truffle risotto, where the truffle was as large as an apple and we shaved slivers from it ourselves.

COOKING

When cooking truffles, the rule is to keep it simple. If you are fortunate enough to find a fresh truffle, slice it, add it to a few beaten eggs and make an omelette, or use it in a simple risotto as above, or sauté it in olive oil with onion and garlic as a sauce for pasta.

Sometimes people pickle the truffles in brandy and then add a few drops of the brandy in their cooking. I have tried this and found that the brandy quickly lost or killed the flavour, but I may have used inferior truffles.

Other recipes team truffles with bacon and cook them *en croûte* or add them to a stuffing to use with game or fowl. This seems to me to be dangerously near gilding the lily. I would prefer my truffle diced, cooked with a little garlic in some olive oil and eaten on toast. No dish could be greater.

SOME FLAVOURINGS FOR VEGETABLE COOKING

AMCHOOR

Amchoor (sometimes spelled amchur) is so wonderfully delicious I am astonished that it is not better known outside Indian cuisine. It is made from unripe mangoes dried in the sun and then finely ground to a powder. With a flavour that is of an intense, almost dark, fruitiness, it is commonly used in northern Indian vegetarian cooking as a souring agent in the same way as tamarind. Generally it is added as a dressing, after the cooking. It is also useful in marinades as it acts as a tenderizing agent: 1 teaspoon of amchoor is equivalent to 3 tablespoons of lemon or lime juice. It is available in most Indian shops, but if you can't get a hold of it an appropriate amount of lime juice will give a little of its effect.

ASAFOETIDA

In its original state, asafoetida is a large bulbous root which grows in the Middle East – mine comes from Afghanistan. It is dried and ground to a powder for use as a flavouring. It is a very ancient spice indeed, much loved in the ancient world.

Its aroma is redolent of garlic that merges into an intense savouriness which can be incredibly powerful – akin to body odours – which some people find utterly repulsive but others adore. In fact, the world seems to be spilt into two camps – asafoetida lovers and haters. I am of the former persuasion and believe this spice to be a huge boon to vegetarian cooking for, in my opinion, it can lift dishes into the gourmet class.

It does, however, lose some of its strength the longer it is stored and the last batch I bought was very fresh and had the kind of pong that could clear not just a room but a football stadium. So this note is a warning: smell before buying and if you don't like the smell leave it alone. Also, if it is very fresh you will probably only need a pinch rather than a half teaspoon or whatever.

SPECIALIST SUPPLIERS

Marshalls, Wisbech, Cambridgeshire PE13 2RF
Excellent for seed potatoes, including La ratte, Pink fir apple and Belle de fontenay.

Suffolk Herbs, Monks Farm, Pantlings Lane, Kelvedon, Essex CO5 9PG
The best source for Chinese and Oriental vegetables, seeds for sprouting, ancient grain crops, herbs and vegetables.

ACKNOWLEDGEMENTS

The Publishers would like to thank the following for supplying vegetables for photography:

Bedfordshire Growers Ltd
Charles Bransden Ltd
Clarissa Dickson-Wright
English Village Nurseries Ltd
Garson Farm
Kew Gardens
Lisdoonan Herbs
Plaxtons of Woodmansey
The Potato Marketing Board
Members of the Roehampton Garden Society
Ryton Organic Gardens
Sapphire Produce
Peter Stovold of John W. Stovold & Sons
Thames Valley Market & Salads Ltd

PICTURE ACKNOWLEDGEMENTS

10 The Image Bank (Cesar Lucas); 11 The Image Bank (Steve Satushek); 27 Tony Stone Images (Michael Busselle); 31 Sutton Seeds; 36 Agence Top (Pierre Hussenot); 59 Jacqui Hurst; 68 Mise au Point (Arnaud Descat); 83 Photos Horticulture; 100 Photos Horticulture; 102 Photos Horticulture; 109 Holt Studios International (Nigel Cattlin); 114 Jacqui Hurst; 121 Robert Harding Picture Library (Nigel Blythe); 129 Eric Crichton; 149 Anthony Blake Photo Library; 183 Zefa Pictures; 217 Denis Hughes-Gilby; 235 Photos Horticulture; 239 Travelpress (Silvio Fiore); 243 Anthony Blake Photo Library; 276 Jerrican (Viard); 277 Agence Top (Jean Noel Reichel).

BIBLIOGRAPHY & LIST OF SOURCES

THE ANCIENT WORLD

Apicius *The Roman Cookery Book* Trans. Barbara Flower and Elisabeth Rosenbaum (Harrap 1958). Critics for years complained of the proliferation of spices in the cuisine which they thought would be bound to conflict, but few of these critics had ever tried the recipes. Then it was pointed out that there were no more spices in Apicius than in Indian cuisine. People who have followed the recipes are enthusiastic about the results.

Athenaeus *The Deipnosophists* 15 vols. Trans. Charles Burton Gulick (7 vols. Loeb Classical Library; London, Heinemann 1961). The original title also means 'The Learned Banquet'. Completed after AD 192, it is written in the symposium form set at a banquet where philosophy, literature, law, medicine and food are discussed. This is our main source of information on what was eaten in the ancient world. Athenaeus cites 1,25O authors, gives the titles of more than 1,000 plays and quotes more than 10,000 lines of verse.

Cato and Varro *On Agriculture* 1 volume. Trans. W. D. Hooper and H. B. Ash. (Loeb Classical Library; Heinemann 1954). Cato (254–149 BC) was brought up on a farm. On Agriculture was written around 160 BC and deals with vine, olive and fruit growing.

Columella *De Re Rustica* 3 vols. Trans. Harrison B. Ash (Harvard University Press 1955). A Spaniard who served as a tribune in the Roman army in Syria about AD 60, Columella composed a treatise on farming, its livestock, fishponds, bees and gardens. He was concerned with the decline of Italian agriculture caused by absentee landlords, the growth of enclosures and dependence on imported food.

Dioscorides (first century AD) *Materia Medica* Trans. M. Wellman (1907–14) now out of print. A Greek who became a physician in the Roman army, his work gives a list of drugs that are natural remedies from vegetable, animal and mineral sources. It became the standard textbook on pharmacy for many centuries.

Herodotus *The Persian Wars* 4 vols. Trans. A. D. Godley. Herodotus (490–425 BC) describes the struggle between Greece and Asia from the time of the 6th century to Xerxes' retreat from Greece (478 BC). Throughout there are many observations on food and customs.

Hippocrates *Hippocratic Corpus* 6 vols., on medicine, disease and diet. Hippocrates was a Greek physician born on the island of Cos about 460 BC. In this work there is much about the effect of certain foods upon the health of the body.

Horace *Odes and Epodes* 1 vol. Trans. C. E. Bennett. Roman poet (65–8 BC).

Horace *Satires, Epistles, Ars Poetica* Trans. H. R. Fairclough.

Martial *Epigrams*. 2 vols. Trans. W. C. A. Ker (Loeb Classical Library; Heinemann 1943) The book includes a collection of elegant couplets designed to accompany gifts of food and drink, or gifts taken home from banquets.

Ovid *Collected Works*. 6 vols. and many translators. There is much evidence that Ovid, Roman poet (43 BC–AD 17), loathed the killing of animals for food and, like Seneca, Plutarch and others, was vegetarian. (See The Heretic's Feast, Colin Spencer, Fourth Estate, 1993)

Palladius, author in the 4th century AD of a Latin treatise on agriculture. All translations now out of print.

Pliny *Natural History* 10 vols. Trans. H. Rackham, W. H. S. Jones and D. E. Eicholz (Loeb Classical Library; Heinemann 1963). A huge and comprehensive work on botany, zoology and the medicinal properties of plants and certain foods. A fount of information.

Xenophon 7 vols. (428–354 BC) Greek historian and disciple of Socrates. *Memorabilia, Apologia and Symposium*, while two others recount his experiences in the Persian wars, where his observation of food as part of the culture is acute.

THE MEDIEVAL WORLD

Albertus Magnus (1200–1280, canonized 1931): the teacher of St Thomas Aquinas and the most prolific writer of his century, he was the only scholar of his age to be called 'the Great'. He translated Aristotle and expounded him but did not endorse him. He wrote seven books on vegetables, *De Vegetabilibus* and twenty-six books on animals, *De Animalbus*. These Latin works have been translated into German but not into English.

The Four Seasons of the House of Cerruti Trans. Judith Spencer (Facts on File Publications, New York 1983). A set of 200 illuminations on herbs, fruits, crops, wild and domesticated food, with accompanying text. A vivid portrait of life in the Po Valley at the end of the fourteenth century, the text carries the tradition of Arabic medicine into Christian Europe.

The Goodman of Paris, trans. Eileen Power (Routledge 1928) *Le Ménagier de Paris*. A treatise on moral and domestic economy by a citizen of Paris was first published in 1846, but it was written in 1393. It was a book written by a wealthy man for the instruction of his young wife. He was at least sixty while she was fifteen when she married him. He tells her how to look after wine, preserve fruits and vegetables, how to prepare water scented with sage, camomile, rosemary or marjoram in which to wash the hands at table, how to cultivate countless vegetables and how to cook them.

THE AGE OF DISCOVERY

Castelvetro, Giacomo *The Fruits, Herbs and Vegetables of Italy*. Trans. Gillian Riley (Viking 1989). Beautifully illustrated, this is a fascinating text. Castelvetro found it difficult to understand the English dislike of vegetables and tried to introduce a better understanding of their cultivation and cooking.

Culpeper, Nicholas (1616–1654): a writer on astrology and medicine. *Culpeper's Complete Herbal* (Omega Books 1985).

Dìaz, Bernal (1492–1581): a Spanish soldier who took part in the conquest of Mexico. He wrote at the end of his life *A True History of the Conquest of New Spain*, a highly readable account which has become a source book of rich information of the foods and diet of the New World.

Evelyn, John (1620–1706) *Acetaria – a discourse of Sallets* (Prospect Books 1982). A great enthusiast for the enjoyment of vegetables, his text contains much information which is applicable today and gives details of other vegetables we have almost forgotten.

Elinor Fettiplace's Receipt Book, ed. Hilary Spurling (Viking/Salamander 1986). An excellent source-book for understanding the cooking of the landed gentry in the Jacobean age.

Gerard, John (1545–1612): published his *Herbal* in 1597. There are many editions, one of the most recent was published by Bracken Books in 1985.

Marquette, Jacques (1637–1675): French Jesuit missionary and explorer. Among his many other journeys, he was the first to travel down the Mississippi river and to report its course.

Martyr, Peter (1457–1526): Chaplain to Ferdinand and Isabella and historian to the Spanish conquests. His collection of 812 letters is valuable source material for the New World.

Muffet, also Moufet and Moffet, Thomas (1553–1604): physician and author.

Murrell, John (c.1630): writer on cookery who improved his knowledge of his art by foreign travel. Published *Two Books of Cookerie* (1638).

Tusser, Thomas (1524–1580): agricultural writer and poet, published *Hundred Good Pointes of Husbandrie* (1557).

WRITERS OF THE PAST

Acton, Eliza *Modern Cookery* (1859). When first published it included a detailed sub-title which said that the book was for private families in which the principles of Baron Liebig have been applied – it was also 650 pages long. An abridged version, *The Best of Eliza Acton* (1974), edited by Elizabeth Ray, is available from Penguin Books.

Adam's Luxury and Eve's Cookery (Prospect Books 1983). First published in 1754, one half is a treatise on kitchen gardening which still has useful hints for today, while the other half is a collection of recipes.

Aubrey, John (1626–1697): antiquary and writer, renowned for his acid, vivid and intimate portraits of his contemporaries. *Brief Lives* (2 vols. 1898, edited Charles Clark.)

Beeton, Mrs *Every Day Cookery* There have been many editions of this book in the last 150 years which have adulterated the original text in a misguided effort to bring it up to date. It began as recipes and housekeeping ideas sent in by readers of Samuel Beeton's *The Englishwoman's Domestic Magazine*, a monthly started in 1852. His young wife then compiled these into a book under her name. The original edition and others from the nineteenth century are invaluable in providing a picture of the Victorian domestic scene.

Boulestin, Marcel *Simple French Cookery for English Homes* (Heinemann 1924). Elizabeth David praised Boulestin for his taste in the manner by which he brought French cooking to England.

Dallas, E. S. *Kettner's Book of the Table, a Manual of Cookery, Practical, Theoretical, Historical* (Centaur Press 1968). First published in 1877, this is a scholarly encyclopaedia of gastronomy, which blends historical research and literary allusions with wit and anecdotes, telling us much about the taste and cuisine of the eighteen-seventies.

Glasse, Hannah *The Art of Cookery Made Plain and Easy* (1747, Prospect Books 1983, a facsimile of the first edition). The best-known cookery book of the eighteenth century, enjoying continuous popularity from the time of its first publication.

Grieve, Mrs M. *A Modern Herbal* (Penguin Books). First published in 1931, with an introduction by Mrs Leyel of Culpeper's, this a marvellous collection of arcane knowledge, remedies and recipes. A source-book for all herbalists and wild food lovers, which gives some understanding of how the majority of humankind gathered food and healed themselves since time immemorial.

The Ladies Companion (1753) is a cookery book owned by Martha Washington.

Layton, T. A. *Choose Your Vegetables* (Duckworth 1963). A good guide, as lively as when it was written, packed with useful information on gardening and cooking.

Leyel, Mrs C. F. *The Gentle Art of Cookery* (Chatto & Windus 1974). First published in 1925, Mrs Leyel's cooking is dominated by a discriminating flair for good things. She was much admired by both Mrs David and Jane Grigson. Mrs Leyel was also the founder of Culpeper's and the Society of Herbalists.

McMahon, Bernard *The American Gardener's Calender, adapted to the Climates and Seasons of the United States* (Philadelphia, 1806).

Miller, Philip (1691–1771): gardener to Chelsea Botanical Gardens, he discovered the method of flowering bulbous plants in bottles filled with water. Published *The Gardener's and Florist's Dictionary* (1724) and *Gardener's Kalender* (1732).

Mortimer, John (1656–1736) *The Whole art of Husbandrie* (1701, sixth edition published in 1761). A popular work.

Ray, John (1627–1705): thought of as the father of natural history in Britain, he published several volumes culminating in his *Historia Plantarum* from 1686–1704.

Sturtevant, E. L. *Edible Plants of the World* (Dover 1972). Written in the late nineteenth century, this is a scholarly comprehensive book on the edible food plants of the world. A lifetime's work that explores the origins and history of cultivated plants.

Turner, William (died 1568): Dean of Wells, physician and botanist, he published his *Herbal* in 1564 which marks the start of scientific botany in England.

CONTEMPORARY COOKERY BOOKS

Bareham, Lindsey *In Praise of the Potato* (Michael Joseph 1989)

Bhumichitr, Vatcharin *Thai Vegetarian Cooking* (Pavilion 1991)

Boxer, Arabella *Book of English Food* (Hodder & Stoughton 1991)

Conte, Anna del *Secrets from an Italian Kitchen* (Corgi Books 1993)

Cost, Bruce *Bruce Cost's Asian Ingredients* (William Morrow N.Y. 1988)

David, Elizabeth *French Provincial Cooking* (Penguin 1964)

Dimbleby, Josceline *The Almost Vegetarian Cook Book* (Sainsbury's)

Gray, Patience *Honey from a Weed* (Prospect Books 1986)

Grigson, Jane *Jane Grigson's Vegetable Book* (Michael Joseph 1978)

Holt, Geraldene *The Gourmet Garden* (Pavilion)

Hom, Ken *Chinese Cookery* (BBC Publications 1984)

Olney, Richard *Simple French Food* (Penguin 1983)

Owen, Sri *Indonesian and Thai Cookery* (Piatkus 1988)

Roden, Claudia *A New Book of Middle Eastern Food* (Penguin 1970)

Sahni, Julie *Classic Indian Vegetarian Cooking* (Dorling Kindersley)

Shaida, Margaret *The Legendary Cuisine of Persia* (Penguin 1994)

So, Yan-kit *Classic Food of China* (Macmillan 1992)

Spry, Constance & Hume, Rosemary *The Constance Spry Cookery Book* (Weidenfeld & Nicolson 1971)

Willan, Anne *French Regional Cooking* (Hutchinson 1981)

A GOOD READ

Griggs, Barbara *The Food Factor* (Viking 1988)

Hartley, Dorothy *Food in England* (Macdonald 1964)

Larkcom, Joy *Oriental Vegetables* (John Murray 1991)

Mabey, Richard *Food for Free* (HarperCollins 1972)

McGee, Harold *On Food and Cooking* (Allen and Unwin 1984)

Phillips, Roger *Wild Food* (Pan Books 1983)

Root, Waverley *Food* (Simon & Schuster)

Stobart, Tom *The Cook's Encyclopaedia* (Batsford 1980)

Toussaint-Samat, Maguelonne *History of Food* Trans. Anthea Bell (Blackwell 1992)

Visser, Margaret *Much Depends on Dinner* (Penguin 1986)

INDEX

Page numbers in *italic* refer to the illustrations